# How to do your Essays Exams & Coursework in Geography and Related Disciplines

Peter G Knight
Tony Parsons

Routledge
Taylor & Francis Group
LONDON AND NEW YORK

Published in 2003 by:
Routledge
2 Park Square, Milton Park, Abingdon, Oxfordshire OX14 4RN
711 Third Avenue, New York, NY 10017, USA

First issued in hardback 2016

*Routledge is an imprint of the Taylor and Francis Group, an informa business*

A catalogue record for this book is available from the British Library

ISBN 13: 978-1-138-17370-5 (hbk)
ISBN 13: 978-0-7487-6676-5 (pbk)

Page make-up by Florence Production Ltd, Stoodleigh, Devon

**Publisher's Note**
The publisher has gone to great lengths to ensure the quality of this reprint but points out that some imperfections in the original may be apparent

# Contents

# Acknowledgements

The authors are grateful to the anonymous reviewers who made comments on this work at various stages in its preparation, and to colleagues who offered advice or information, particularly Ian Stimpson, Glyn Williams and Phil Kivell. The origin of the checklist at the end of Chapter 16 is uncertain, and we apologize if it has an author whom we have inadvertently failed to acknowledge. PGK would also like to thank Tony Trott, Peter Bull, Judith Pallot, and his Mom for helping to hone his essay-writing skills.

# Preface

**If you are a student looking for help writing an essay, if you are producing some other kind of coursework assignment, or if you are facing an exam, then skip this section and move directly to Chapter 1. This section is intended for people who want to know *about* the book, rather than people who urgently need to *use* the book!**

This book is written for students who need help doing their coursework and exams. It is written primarily for students taking courses in geography, but the principles that we emphasize apply equally to coursework and exams in many other subjects. We focus primarily on the skills and techniques that apply to essay writing and that can be easily transferred to other forms of structured presentation such as posters, talks and web pages. We will not explicitly consider laboratory reports, short-answer tests or problem sheets, because although these do feature in many geography courses they do not seem to present students with the same types of difficulties that they face in producing more extended pieces of work such as essays. The traditional approach to assessment by a combination of coursework essays and exam essays has been extended in recent years to include a wide variety of modes of assessment. Although end-of-course exams are still dominated by essay questions, geography students can now expect their coursework assignments to include poster presentations, talks and screen-based electronic submissions such as web pages or PowerPoint presentations.

The basis of the book is that all of these different types of work are based on clear communication of well-supported responses to the questions or tasks that have been set. This applies equally to a good essay, a good talk or a good web page. The details of how this target is best achieved in each medium may vary, but the essence of 'good work' is the same in each case. The aim of this book is to reinforce the essential principles of good work in the students' approach to coursework and exams, and to help students to see ways in which they can apply these principles in the different situations that they will encounter in their courses.

Most institutions will certainly provide their own guidelines for the preparation of coursework. For the most part we anticipate that the advice in this book will complement and supplement these guidelines, and not contradict them, because we have aimed to produce the kind of advice that most of our colleagues in other institutions would support. However, as there will inevitably be points of detail where specific institutional guidelines may part company from our advice, we are careful to explain to students that their institution's guidelines, whether we appear to endorse them or not, will be the guidelines that determine the outcome of their assessments.

We have called the book ' . . . in Geography and Related Disciplines' because, although we are aware that much of what we say will apply equally well to other disciplines, we have written with geography students specifically in mind, and have written from our own perspective as geographers involved in teaching geography. Our case studies and examples all relate to topics in geography, and our references to administrative or academic contexts such as Subject Centres and Benchmark Statements are all based on geography.

The book is written explicitly for students, rather than their tutors, and is written with a style and structure intended to make the book user-friendly to students who are struggling to improve their work. The book might usefully be introduced right at the start of a student's career, perhaps in the context of a study-skills class or small-group tutorial. While it might not be a 'course text' for any specific module, we hope that it will be of value to students across their geography course and throughout the length of their career. We have written the book as a companion volume to our earlier *How to Do Your Dissertation in Geography and Related Disciplines* and we have followed a number of precedents from that book that have proved popular with users. For example, the book is organized in the form of a series of questions of the type that students might be asking themselves at each stage of their work; each chapter includes a brief summary of key points; and we have included contributions from 'our friend Eric' to provide a counterpoint to the dull rhetoric of the old fogies who wrote the bulk of the text.

# What is this book for, and how should I use it?

## *Chapter summary*

The aim of this book is to help you to write better essays and to get better marks. It is designed to help you with both your coursework and your exams, and also to help you with other types of presentation, like posters, talks and web pages. Using this book will develop your skills and improve your confidence, leading to better work and better grades. When you first use the book, look quickly through it to see what kinds of things it covers and to gather the general principles it puts forward. Then use it as a step-by-step guide for your next few pieces of work. Afterwards, use it as a reference guide whenever you have problems with specific assignments, and as a revision and preparation aid before exams.

## What is this book for?

The aim of this book is to help students in geography and related disciplines to write better essays and to get better marks. The book also covers other types of assignment like posters, talks and web pages, and it deals with both coursework and exams. If you are doing a course where you have to write essays, put together talks or posters, or make any other kind of structured presentation, then this book is written specifically to help you.

We've written this book because most students don't do as well as they could do with their assignments. Most students could do much better work, and get much higher marks, if they just followed a few simple guidelines.

**We want you to do the best work you can, so we've written this book to help.**

Some courses require students to do most of their coursework in the form of essays, while others require you to produce a variety of essays, posters, talks, web pages and other forms of work. The basic principles behind all these types of work are the same, and as well as explaining the ins and outs of writing essays, this book will also explain how to transfer your essay-writing skills to other types of presentation. We will also deal with the differences between the kind of work you can do when you have a whole semester to produce a report and the kind of work you can do when you have just a couple of hours in an examination room. Again, the same principles apply to both situations, but you have to learn to adapt your skills to get the most out of both types of test. This book will show you how.

## How should I use this book?

You will have noticed by now that we've arranged the book in the form of a series of answers to the specific questions that we think you will be asking as you read it.

**The book is a bit like a workshop manual or a recipe book.**

You can dip into it to look up answers to specific questions, or you can work through it section by section as you tackle your coursework projects. We have written it with the expectation that you will do a little bit of both.

We suggest that you should start off by taking a quick skim through the book (especially the contents page and the chapter summaries) just to see what kinds of things we talk about. This will give you a clue as to when you might want to come back to it in moments of panic (no, don't panic yet). Next, if you are a wise and right-thinking student you will at some point early in your career (before things become urgent) read the book carefully from cover to cover. Trust us, it will stand you in good stead. After you've done that, we suggest that you take the book for a serious test drive by using it step by step as you do your next piece of coursework. When your assignment is first set, use Chapters 5 and 6. As you start to assemble your materials and plan the work, use Chapters 7 and 8, and so on until just before you submit the work, when you should look at Chapter 15. Of course, different sections might be more or less appropriate to specific assignments that you do, but once you know your way around the book you will quickly be able to choose which bits offer most relief in times of pain.

The reader we've had in mind while writing this book is a student wanting help with an essay. It's written for you, not for your tutors or for your granny. Therefore we've written in a way that is intended to make the book as easy as possible for you to use. We've written in a pretty informal style, but we should point out right away that the style we've used to write this book is not the style you should normally use to write your essays! You always have to choose a style appropriate to your particular piece of work. We'll go into that in more detail later (Chapter 8).

Of course, this book is not your only source of advice and assistance. Your institution will almost certainly produce a set of guidelines of its own that explain to you the regulations governing your coursework and the nuts and bolts of what your tutors expect of you. In the end it is they, not we, who will be marking your essays, so you should always seek their advice. If their advice seems to conflict with ours, do what you think best. It is you, not we, who will have to defend your work in a tutorial or in a post-exam viva.

**Always begin by reading your institution's guidelines and your tutor's instructions.**

## What types of assignment are (and are not) covered in this book?

This book specifically covers essays, talks, posters, professional reports, newspaper articles, press releases, abstracts and computer-based presentations such as web pages. The book does not cover laboratory classes, short-answer exercises or problem sheets, and it does not cover dissertations or project reports.

We have chosen to cover those particular types of assignment because they form the core of the independent work that most geography students are asked to submit as coursework and in exams, but there are not many books available that provide specific advice on how to do them – at least, not friendly books like this one. We are able to cover all these types of assignment in one book because the same basic principles apply to all of them. For example, if you learn how to do a good essay, it is easy to transfer that skill to giving talks or preparing posters and web pages. These are all structured presentations that follow the same basic rules, but the different media used to convey the material affect how you apply those rules. We don't cover lab-class reports or problem sheets because they follow different rules, and we don't cover dissertations because we have already written another book about dissertations (*How to Do Your Dissertation in Geography and Related Disciplines*), which will be much more use to you than this one will if doing a dissertation is your immediate problem.

## Our friend Eric

Whenever you're struggling with a piece of work it's useful to have a friend to bounce ideas off, to share problems with, and maybe even enlist to read through a rough draft to see how it sounds. It's useful to have somebody to give an alternative point of view. In this book, when we need someone like that, we use our friend Eric. Eric can be a good friend to have around. He can suggest alternative ways of doing things, he can get you out of bed in the afternoon when your 5 o'clock deadline is approaching and, when he has something really important to say, he can send you a note:

Eric says: *You need all the help you can get, but in the end you have to make up your own mind and do your own work.*

## Conclusion to Chapter 1

- This book will help you to do better essays, exams and other coursework.
- You can use it either for reference or as a step-by-step workshop manual.
- You should use it alongside your tutor's instructions and advice.

2

# Why are there so many different types of assignment?

## Chapter summary

**Assessments of your work are intended to allow you, your tutors, and potential employers to judge your progress and achievements. Different types of assignment allow you to demonstrate different skills and abilities, and geography students can expect to be given a range of types of work to do. Although your assignments will involve different skills, they are all based on similar fundamental principles. This book will help you to identify those principles and to recognize how they should be applied to different tasks.**

## Why have assessment at all?

Assessment serves several different purposes. First, it lets you know how you are getting on throughout your studies. The marks you get week by week, or semester by semester, are a barometer of your progress that you can use to steer your way towards your final grade. Second, assessment tells your tutors how you are getting on. This is useful to them as they try to give you the most constructive and helpful advice that they can, and it also provides evidence that can, and usually will, be chalked up against your name as part of some final course grade. Third, assessment, especially the cumulative final score or degree result, provides a label that will be attached to you and will follow you out of university and into the big bad world of job-hunting, or whatever you choose to do next. To your dying day it will tell you, and everyone else, how well you did at university.

Eric says: *You ought to try to do the best you can. You'll regret it later if you don't.*

## What does assessment actually measure?

The marks that you get for your work are supposed to indicate how well you have done. Tutors, examiners and potential future employers look at them as a measure of your achievement. It is important for you, and for them, to realize what is actually being assessed. This has been made easier by the production of precise formal guidelines that describe exactly what is expected of students graduating with different levels of qualification. These guidelines have been produced by the Quality Assurance Agency for Higher Education (QAA) within a document called 'Framework for Higher Education Qualifications in England, Wales and Northern Ireland'. For students graduating at the level of a bachelor's degree with honours, the QAA identify a number of things that students should be able to do. We've put a version of this list in Box 2.1 so that you can refer back to it when you need to be reminded what you are trying to convince your examiners of.

Study this list. This is what is expected of you. This is what your examiners will be looking for. This is what is being assessed.

### Box 2.1 What can be expected of a graduate

*Based on the descriptor for the qualification of bachelor's degree with honours, from the Framework for Higher Education Qualifications in England, Wales and Northern Ireland produced by the Quality Assurance Agency for Higher Education.*

Students should be able to:

- evaluate evidence, arguments and assumptions, to reach sound judgements, and to communicate effectively;
- devise and sustain arguments, and/or to solve problems, using ideas and techniques, some of which are at the forefront of a discipline;
- describe and comment upon particular aspects of current research, or equivalent advanced scholarship, in the discipline;
- apply the methods and techniques that they have learned to review, consolidate, extend and apply their knowledge and understanding; and to carry out projects;
- critically evaluate arguments, assumptions, abstract concepts and data (that may be incomplete); to formulate judgements, and to frame appropriate questions to achieve a solution – or identify a range of solutions – to a problem;
- communicate information, ideas, problems and solutions to both specialist and non-specialist audiences.

## Why use different types of assessment?

Whichever institution you study at, if you do a degree in geography or a similar discipline you can expect to encounter many different types of assessment. From your point of view, this is actually a good thing. It means if there is a particular type of assignment that you don't like you can be sure that your final mark won't depend on that type of work alone. This is one of the reasons that you are given different types of assignment, but there is more to it than that.

Geography is a broad subject, not only in terms of the range of topics that you will be expected to study, but also in terms of the types of skills or abilities that you will be expected to learn. That was clear from Box 2.1. These skills include some that are expected to be of value to you long after you have finished your geography degree. Box 2.2 is an extract from the 'Benchmark Statement' for geography published by the Quality Assurance Agency for Higher Education in 2000, and it shows how much importance is given to the issue of skills training in modern geography.

## Box 2.2 Student skills

*Extract from section 4 'Student skills, abilities and attributes' of the Quality Assurance Agency Benchmark Statement for Geography, 2000.*

'Many geography degree programmes are now at the forefront of policies to furnish students with skills that are valued in the world of work and provide the basis for lifelong learning. Students therefore learn "through" geography in addition to learning "about" geography. The attention given to skills, both discipline-specific and generic, is intended to improve students' academic performance, enhance their career prospects, enable them as citizens to make a full contribution to the wider community and give them the flexibility required to adapt to new developments and opportunities in a rapidly changing world.'

The main reason that you find yourself having to do all sorts of different types of coursework and exams is that you are being taught, and being tested in, all sorts of different skills. Different types of assignment allow you to demonstrate your competence in different areas. Or, to see it from another perspective, it allows your examiners to see how good you are at doing different things.

In the past, most geography courses employed a wide range of assessment types as a matter of course because it was clearly good practice. Recently this good practice has been formalized and to some extent standardized by the production of a 'Benchmark Statement' by the QAA. This Benchmark describes the sort of things that a geography programme should be expected to offer and achieve, and within it is a description of the type of assessments that will usually be included as part of the programme. We've put (a version of) this description here, in Box 2.3. It should help to explain why your tutors keep asking you to do so many different things.

Box 2.3 Assessment in geography

*From the Subject Benchmark Statement for Geography, produced by the Quality Assurance Agency for Higher Education.*

Students should be permitted to demonstrate their full range of abilities and skills, with institutions providing a mix of assessment methods that are, overall, equally accessible to students from varying educational backgrounds and in different learning situations. Students of geography are therefore likely to encounter most of the following assessment methods in their degree: unseen examinations with a range of types of questions/tasks; dissertations and projects (and proposals for these); practical work (in the field, scientific laboratories, specialist C&IT work and quantitative and qualitative analyses); essays of varying lengths; reports; oral presentations; posters; press releases; annotated bibliographies; objective tests (perhaps through computer-based assessment); internet-based assignments; work-based assessments; and teamwork of varying kinds. Geography has been innovative in the development both of assessment of these types of learning and of ideas of equity and consistency of standards.

## Is all assessment equally important?

Some pieces of your assessed work will count towards your final degree classification. These are clearly important! Some pieces, however, will not. So if they don't count, why should you bother to make any effort? Why has your tutor asked you to do them at all? It's not just to keep you busy, when you could be down the pub. After all, if you've been set the coursework, it means your tutor is going to have to mark it when he or she could have been down the pub.

Believe it or not, if your tutor has set you a piece of coursework it's because it is in your best interests. And it's in your best interests to do it well.

Generally, if the assessment doesn't count, it's because the coursework has been given to you as an opportunity to practise. It's your free go. So don't waste it. If you do well, it means that when you come to the real thing you'll have the confidence that you can do it again. If you do badly your tutor will be there to help. No matter how well you do, your tutor will look for ways to help you improve your work, and is likely to make comments according to how well you have done. So if your piece of work is worth 40% your tutor will be saying things that will help you get 50% or 60%. If you get 70% your tutor will be suggesting things to help you get 80% or 90%. If you were capable of getting 70% but just didn't bother to make any effort, your tutor will be telling you things that are no help at all, and you'll never know what it was that could get you 90%.

People often use the terms 'formative assessment' and 'summative assessment'. Formative assessment is where you get feedback that will help you do better next time. In summative assessment there is no next time. So make sure you get all the benefit you can out of formative assessment, whether it counts or not. If you do well in your formative assessments you should be more confident that you can do well in your summative ones.

Eric says: *Confidence is an important weapon in your armoury, and your tutors will try to help you to develop it.*

## How will I learn how to do all these different things?

So far all we've done is tell you what you already know: that you are expected to do lots of different types of work within your geography course. What we need to move on to is explaining how to cope with them all. What you need to know is how to do well in all these types of work. To recap, we've established that you will need to be competent in the following:

- essays;
- posters, and other forms of written communication;
- web pages and other on-screen presentations;
- oral presentations (talks);
- group work.

Different institutions offer different amounts of training in these different skills. Typically, students will get a little explicit advice, but will learn mainly from tutors' feedback on the weaknesses in the work after it's been marked. What we will try to do in this book is give you some important pointers *before* you lose marks by making the mistakes in your assessed work.

Fortunately for us, the same basic principles apply to all these different types of coursework. We will begin by identifying those basic principles, then we will explore how to apply them by looking in detail at essay writing, and then we will see how you have to change your approach and do things just a little bit differently to do well in other types of assignment.

## What basic principles apply to doing well in all types of assignment?

All the different types of work that we consider in this book are based on students being set a specific task. The task may be to describe something, to explain something, to compare several things, or to complete some combination of these or other tasks. The key point is that the tutor or examiner will be asking you to do something quite specific. Your job is to do what is asked of you, to do it clearly, and in doing so to demonstrate your competence or expertise both in the subject and in the mode of presentation.

**The most important thing is to do what's being asked of you.**

This may sound too obvious to be useful, but most things about doing good work are obvious when you write them down. Here's an example. Suppose you are given an essay with a title in the form '*Explain the difference between A and B . . .*'. It would be easy to write an essay that describes A and B very clearly but doesn't actually explain why they are different. It would also be easy to write an essay that explains the difference between A and C. Either of these could be good essays in their own right, but both would get you a fail mark if you produced them in answer to the question that was set here.

Eric says: *Show them how much you know, but be careful to show them that you know which bits are relevant and which bits should be saved for another day!*

Once you have made sure that you are doing what has actually been asked of you, your second priority is to make sure that you do it clearly. For a start, you have to make sure that they can see that you've done what they asked. Even if you thought you had answered the question, you won't get credit for it if the tutor can't tell that you've answered the question. Producing a clear answer involves several different skills, ranging from logical organization of the whole piece to technicalities of legible presentation in whatever medium (written, spoken, on-line) that you are using.

Eric says: *If you've gone to the trouble of doing the work, you might as well make sure they can tell!*

Finally, if you have done what was asked of you, and done so with sufficient clarity that your work can be understood and appreciated, you must make sure that you have taken the opportunity to demonstrate your expertise in the subject. This can be achieved not only by giving a sensible and well-reasoned answer to the question, but also by including appropriate amounts of relevant information.

Eric says: *When they set you work, figure out exactly what they want, and give them just that. Nothing else will do, however good it might be!*

These three key goals – relevance, clarity and detail – apply to all the work you will do. What we need to think about now is how to achieve them in different types of assignment. Read on!

## Conclusion to Chapter 2

- It isn't just malice that makes your tutors give you so many different types of assignment. They are helping you to develop different skills, and assessing how well you put those different skills into practice.
- Good practice in geography, which can be illustrated for example by guidelines such as QAA Benchmarking Statements, specifies that students should experience a range of types of assessment.
- Different types of assessment exercise different skills, but are all based on the same fundamental principles of clear explanation in different media. This book will help you to develop those principles in the context of different types of assignment.

# 3 What are essays, and why do I have to write them?

## *Chapter summary*

Essays are both a learning tool and a means of testing. As a means of testing, essays provide an all-round guide to your academic abilities. Among other things, they tell the reader what you know, how well you can organize ideas, and how well you can communicate. As a learning tool, essay writing forces you to research and collate material, mentally manipulate a question or problem, structure your own thoughts, prioritize and argue points in a logical manner, communicate effectively, write syntheses of complex material, and deal with many other intellectual and practical issues. When revising for exams, writing essays is a useful way of organizing your knowledge, and essays that you wrote during your course are useful sources of material for revising from. Essays are good for you!

## What is an essay?

An essay is a piece of writing that allows you to put forward your knowledge and understanding of a topic at length and in depth, with the opportunity to consider alternative viewpoints, to express your own opinions, and to develop a balanced, reasoned and logically structured argument.

> An essay is not just a list of facts, figures and assertions. It is a means of communication.

## Why do we write essays as part of our degree studies?

Essays serve two basic functions. They help students to learn, and they help tutors to assess students. Sometimes it is clear that an essay is fulfilling just one of these functions. For example, in your final exam the essays you write are intended simply to give you an opportunity to impress the examiner and earn a good mark. In other situations, however, an essay will be doing both jobs at once.

For example, many of your term-time essays will be assessed by your tutors to generate marks that count towards your assessment, but at the same time they are intended as a learning aid.

Both the process of writing an essay and the process of receiving feedback (marks and comments) from your tutor are intended to help you not only to write better essays but also to develop a range of subject-specific and transferable skills. Some of you in your future careers will have to complete pieces of writing that allow you to put forward your knowledge and understanding of a topic at length and in depth, so essay writing is in itself an important skill. Even if you never have to write another essay or report in your life, the ability to gather, organize and communicate information is an important part of what justifies getting a university degree. You need to acquire this skill, and demonstrate that you can apply it, in order to succeed in your course. In addition, for those of you who resent any activity that does not immediately and directly increase your stock of factual geographical knowledge, there is no better way to learn about a topic than to research and write an essay on it.

It is important that you appreciate the significance of essay writing. Most students have to do a lot of it, and doing it well will make a big difference to your final result. It will help you to treat your essays with the respect they deserve if you are absolutely clear both about what benefits you are getting by writing them and about why your tutors are asking you to write them. That's why we've included the next two sections in this chapter.

Eric says: *You always do things better if you understand what you're doing them for!*

## How do students benefit from writing essays?

Essay writing forces you to research and collate material, to manipulate a question or problem in your mind, to prioritize and argue points in a logical manner, and to communicate effectively within the conventions of the discipline. In other words it forces you to find stuff out, to think about it, and to tell someone (the reader) what you make of it.

Eric says: *Writing essays is good for you.*

Essay writing really cuts directly to the heart of your learning. In order to write a good essay you have to do all the things that your course is about: learning, thinking and communicating. If you need to be convinced, check out Box 3.1.

### Box 3.1 How essays help you

*Essays help you to learn material about course topics:*

- gathering new material from journals, books and other sources;
- revising and assimilating lecture notes by trawling them for relevant material;
- increasing broad understanding of the topic by spending time thinking about how different items of material relate to each other;
- developing insight into specific aspects of a topic by assessing evidence for and against particular points of view.

*Essays help you to learn and practise skills:*

- researching and collating information from a variety of sources;
- assessing the merits of conflicting information and contradictory viewpoints;
- using evidence to argue for and against particular hypotheses or points of view;
- writing clearly, concisely and convincingly;
- adopting the conventions of professional scholarly writing (e.g. reference citations) in particular disciplines (e.g. natural sciences, social sciences).

## What do tutors learn by reading students' essays?

Reading an essay tells a tutor an awful lot about the student who wrote it. For a start it reveals very quickly whether the student knows much about the topic, whether they have put much effort into their out-of-class assignment, and whether they know the basic techniques and conventions of essay writing. An experienced tutor can usually tell at a glance, without having even to read the essay, whether the student who wrote it is good, bad or average. Your essay will enable your tutor to see what you know, how well you can organize ideas, and how well you can communicate on paper. In other words, your essay is an all-round guide to your academic abilities.

Eric says: *Your essay is the evidence that your tutor will use to judge you, and it is up to you to make yourself look good.*

For a tutor's eye view of what essays are really all about, look at Box 3.2.

### Box 3.2 What your essay tells your tutor

*Your essay will give your tutor, or your examiner, answers to all these questions:*

- Do you understand the question or instruction in the assignment?
- Can you identify the key issues raised by the assignment?
- Can you place those issues into the broader framework of the subject?
- How much do you know about the topic?
- Can you use your knowledge and understanding to answer the question?
- Have you understood the course material on which the assignment is based?
- Have you worked independently beyond the basic course material?
- Are you up-to-date with relevant literature?
- Can you tell which of the things you know are relevant and which are not?
- Can you put together a logical and coherent argument to make your point?
- Can you provide evidence to support your point?
- Can you correctly cite the sources of your evidence?
- Are you widely read and well informed?
- Are you a good communicator?
- Can you organize material in a logical fashion?
- Can you structure your argument and your evidence into a coherent essay?
- Can you write clearly and concisely in good English?
- Can you write in a formal or technical style appropriate to the assignment?
- Can you use diagrams, quotations and other illustrations appropriately?
- Can you set out text, tables and diagrams clearly and effectively?
- Can you present your essay professionally?
- Can you work to a deadline and follow instructions?
- Are you working hard and trying to do well on your course?

## Are there differences between essays in different subjects?

Most geography students take courses in both human and physical geography, and tutors are often asked whether the essays in each branch of the subject should be approached differently. Readers of this book may be studying other subjects and may wonder whether the guidelines that we give here apply equally in history, biology, music, or any other subject. The answer is that there are differences of detail between subjects, but the fundamentals apply broadly between subjects.

Different subjects, or even branches within the same subject, sometimes adopt different conventions. For example, many geography courses (and most books and journals in geography) favour the 'Harvard' system of referencing, where a cited author's surname and the date of publication are given in the text, with the full bibliographic details given in a reference list at the end of the essay. However, some other subjects, and some publications in geography, use the 'Vancouver' or footnote system, where citations are marked by a number like this [(1)] in the text, and details are given in a note at the foot of the page or at the end of the essay. (We discuss the relative merits of these systems in Chapter 12.) It is important that you use the system specified by your tutor. Failure to do so will suggest that you can't follow simple instructions, and you may be penalized.

**Your ability to follow recognized conventions within your discipline or sub-discipline is one of the things that you will be judged on.**

The similarities in essay writing between disciplines are more fundamental than the differences. In every case the student must provide a clear, sound response to the specific question asked or task set by the title. In every case students must demonstrate their knowledge, understanding and ability to communicate. In every case students must adhere to the regulations and conventions of their course and their subject discipline. These similarities provide us with the raw material to identify what makes a good essay (or a bad essay) in any situation; they can be applied to other types of coursework like posters, talks and web pages; and they therefore form the basis of the remaining chapters of this book.

## Conclusion to Chapter 3

- Writing essays should help you to learn, think about and understand your course material. You should use essays to help you to study.
- Essays provide your tutors with a way of assessing your progress and assigning a mark to your achievements.
- Essay writing involves a wide range of skills, and requires a certain amount of factual knowledge, so your ability to write essays gives a broad impression of your overall academic progress. Many of the same skills apply to different types of coursework.

# What will the examiners be looking for when they mark my essay?

4

### *Chapter summary*

Your examiners will be trying to judge whether you have a good knowledge and understanding of the subject, whether you can organize your knowledge to provide a reasoned and well-supported response to a specific question, and whether you can communicate effectively following the conventions of writing in your discipline. Most essays are marked according to a strict set of criteria that are written down as a guide for markers. You should study the guide for markers that your tutors use, as this will tell you exactly what your essay will be judged on. You will not be penalized just because you disagree with your tutor in an essay, but you will be penalized if your opinion is not supported by a well-reasoned argument and good evidence. Essays written in different situations (such as exams and coursework) can be marked in different ways, and you need to learn to interpret the feedback that you get from your tutors.

## The basics

The mark that you get for an essay depends on whether the person who marks it thinks it's any good. To convince the marker that your essay deserves a good mark you have to show two things:

1. essay-writing skills;
2. knowledge and understanding of the subject.

**Getting a good mark for your essay isn't rocket science. It's common sense and hard work.**

## How will the marker judge my essay?

In most courses, essays are judged against a standard set of criteria used by all the different markers throughout the course. The marking criteria are usually published in the course handbook, or provided to students in a handout or on the course web page. If you have not had the marking criteria for your assignments explained to you, you should ask your tutor about them. The people who mark your essays will be doing so with these criteria in mind, so you need to know what they are.

Eric says: *Your tutor will judge your essay on the basis of whether it is any good or not. Before you start, you need to know what the tutor will think is 'good'.*

It's important that you know what you are aiming for when you write your essay. Check your course handbooks to see how marking works in your institution.

## What exactly is the examiner looking for?

The person who marks your essay is trying to find out a number of things about you, and your essay provides the evidence. Box 3.2 listed some of the questions that your marker would get answers to by looking at your essay. For each one of these questions to which the examiner can answer 'yes' you will score a higher mark. For every 'no' you will score lower. Take another look at Box 3.2.

With a list of questions like the one in Box 3.2 you can judge your own essay before it ever goes to the examiner. Ask those questions about yourself, and think about what answers a reader would reach from reading your essay. As long as you understand the criteria on which your work is being judged you are in a good position to make sure that the work is excellent before you hand it in. As we progress through this book we will explore the issues raised by each of the questions in this list and show how you can make sure that your essay encourages the marker to answer 'yes'.

## How are these criteria used to grade essays?

It is a simple task to produce a list of the criteria by which your essays will be judged. It is more difficult to say exactly how many marks are given for each of these criteria, and how the examiners arrive at a final mark or grade. Allocating grades to essays can be a complicated business.

Some institutions may use a numerical approach. For example, an essay may be given a mark out of 10 for each of a range of different criteria, and the grade will be based on the aggregate mark (Box 4.1). However, this system can get complicated because essays may have to reach a certain level in each category, as well as achieving a specified total. For example, in the scheme in Box 4.1 you need a total of 40 points to achieve a first-class grade, but it's no good getting four 10s and a zero. If you score a zero in any category you will be limited to a grade D, even if all your other categories are perfect. If you score less than 3 out of 10 in any of the five categories, your grade will be limited to a C. Find out what scheme your tutors will be following when they mark your essays. If you don't like the scheme that your institution uses, try making up one of your own. In fact, that's quite a good tutorial exercise.

## Box 4.1 A mark sheet for a numerically-based grading system

| | Tutor's comments | Mark (out of10) |
|---|---|---|
| Content | | |
| Structure | | |
| Evidence | | |
| Presentation | | |
| English | | |
| | Total | |

**Grade requirements:**

**A**: Total of 40 or more, with no mark lower than 7.

**B**: Total of 30 or more, with no more than one mark lower than 5 and no mark lower than 3.

**C**: Total of 20 or more, with no more than three marks lower than 4 and no mark lower than 2.

**D**: Total of 10 or more, with no more than one mark of 0.

**E**: Total below 10.

| | |
|---|---|
| Grade | |

Many institutions prefer a less numerically-based approach than the one we've just described, and design guidelines based on qualitative descriptions of essays. Box 4.2 gives you an example of what this kind of marking guide might look like. You will see that it covers the same issues as those in Boxes 3.2 and 4.1, but identifies critical qualitative thresholds for key criteria. For example, a first-class answer cannot contain any significant factual errors or misconceptions; an upper second-class answer may contain some factual errors or misconceptions; a lower second may reveal shortcomings in knowledge or understanding; a third-class answer may reveal limited knowledge and understanding; and a fail may be seriously flawed with errors, omissions and irrelevant material. You will also see that the criteria are cumulative. Factual correctness is necessary for a first-class mark, but it is not sufficient. The essay must also be presented to a high standard and show evidence of independent thought. Even if your essay is factually correct, Box 4.2 indicates that if the essay lacks evidence of independent thought it will not score higher than a lower second-class mark. If well designed, this type of marking guide can provide very thorough support for both students and markers.

## Box 4.2 An example of a qualitative marking guide for geography essays

| Class | % | Criteria |
|---|---|---|
| I | 90–100 | A virtually faultless answer that could not be bettered under the circumstances. Must show detailed knowledge of the subject, deep understanding of all the issues raised by the assignment, original insights based on independent thought, and rigorous and logical argument based on evidence drawn from a wide range of high-level source material. Must be organized and presented to an impeccable standard in the style appropriate to the assignment, without errors of style, language or presentation. Must not contain any factual errors or misconceptions. |
| | 80–89 | An exceptional answer with very few weaknesses. Must show very detailed knowledge of the subject, deep understanding of the issues raised by the assignment, original insights based on independent thought, and rigorous and logical argument based on evidence drawn from a wide range of high-level source material. Must be organized and presented to a high standard in the style appropriate to the assignment, with no significant errors of style, language or presentation. Must not contain any factual errors or misconceptions. |
| | 70–79 | An excellent answer with no serious weaknesses. Must show detailed knowledge of the subject, deep understanding of the issues raised by the assignment, original insights based on independent thought, and rigorous and logical argument based on evidence drawn from a wide range of high-level source material. Must be organized and presented to a high standard in the style appropriate to the assignment, with few errors of style, language or presentation, and no significant factual errors or misconceptions. |
| 2.i | 65–69 | A very good answer, but with some weaknesses. Must demonstrate knowledge, understanding and independent thought, and must be organized and presented to a good standard in the style appropriate to the assignment. Must use logical argument based on evidence drawn from high-level source material. May lack the insight, detail or organization of a first-class answer, or may contain some minor factual errors or misconceptions. |

*Contd*

| | | |
|---|---|---|
| | 60–64 | A good answer, but with significant weaknesses in content, organization or presentation. Must demonstrate knowledge, understanding and independent thought, and must be presented in the style appropriate to the assignment. Must use logical argument based on evidence drawn from high-level source material. May have shortcomings in insight, detail or organization, or may contain some factual errors or misconceptions. |
| ii | 55–59 | A reasonable answer, but with significant weaknesses in content, organization or presentation. Must recognize the key issues raised by the assignment, and provide a relevant answer, but may reveal some shortcomings in knowledge or understanding. Must demonstrate knowledge and understanding of course material, and must use evidence to support arguments, but may lack evidence of independent thought. |
| | 50–54 | A reasonable answer, but with significant weaknesses in content, organization or presentation. Must recognise the key issues raised by the assignment, and provide a relevant answer, but may reveal shortcomings in knowledge or understanding, and may include errors, omissions and some irrelevant material. Must demonstrate basic knowledge and understanding of course material, and must use evidence to support arguments, but may lack evidence of independent thought. |
| | 45–49 | A poor answer that reveals limited knowledge and understanding of the subject but nevertheless recognizes most of the main issues raised by the assignment. Must provide a relevant answer, but may include a few serious errors, omissions or irrelevant material. May lack any evidence of independent thought. Arguments may lack evidence, and there may be no indication of independent reading. May not be presented to a high standard in the style appropriate to the assignment. |
| | 40–44 | A poor answer that reveals limited knowledge and understanding of the subject but nevertheless recognizes some of the main issues raised by the assignment. Must provide a relevant answer, but may include serious errors, omissions or irrelevant material. May lack any evidence of independent thought. Arguments may lack evidence, and there may be no indication of independent reading. May not be presented to a high standard in the style appropriate to the assignment. |

*Contd*

| | | |
|---|---|---|
| Pass | 35–40 | An unsatisfactory answer, but one which shows some understanding of the relevant issues, and is presented to a standard that indicates a basic appreciation of the style appropriate to the assignment. |
| Fail | 30–34 | An unsatisfactory answer, but one which indicates an understanding of the assignment. Must include some relevant material, but may be seriously flawed with errors, omissions and irrelevant material. |
| | 20–29 | An unsatisfactory answer that indicates little understanding of the assignment, and may contain little relevant, accurate or appropriate material. |
| | 10–19 | An unsatisfactory answer that shows only the most limited evidence of having benefited from the course. |
| | 0–9 | No answer given; or an answer which is completely irrelevant or fundamentally wrong. This range of marks can also be applied to work that is plagiarized. |

In most institutions, each department or course team will produce a separate set of criteria for their particular needs. Geographers, for example, might have slightly different criteria from molecular biologists or musicians. There might also be different guidelines for marking different types of work. For example, there are things that would be judged in a spoken presentation that do not apply to an essay. Box 4.3 is an example of an actual markers' guide from a UK university covering both essays and spoken presentations. We will look in more detail at the differences between different types of work later (Chapters 17–20).

What you need to do now is find out what marking guidelines are used in your course. You should get hold of a copy, and you should refer to it as you work your way through this book and through each of your assignments.

## Will all my essays be marked in the same way?

How the marker approaches your essay depends on whether it's an exam essay, a coursework essay that counts towards your final assessment, or a coursework essay that does not count towards a final assessment. Essays written for different purposes will be handled in different ways.

For an exam essay, the marker will simply be concerned to assign a fair and accurate mark to represent the quality of your work. The criteria that your essay will be judged against should have been made clear to you before you sat the exam, and will certainly be clear to the examiner, as we discussed in the previous section. The examiner will not be concerned with explaining the mark to you, or helping you to learn from the exam experience, and if the marker writes on the essay, the notes will only be intended for the other examiners who check the marking.

## Box 4.3 The qualitative marking scheme used by one UK University Department of Geography for projects/essays and oral presentations

| CLASS or GRADE | MARK RANGE | DESCRIPTIVE EQUIVALENT FOR: | |
|---|---|---|---|
| | | Projects/Essays | Oral presentations |
| FIRST-CLASS HONOURS | 91–100 | Worthy of retaining for reference | Content:<br>Strong and clear structure<br>Literature extensive and fully integrated<br>Very well developed argument<br>Original thinking/contribution<br>Strong critical appreciation of topic<br>Data/information perceptively used<br>Very good and effective effort<br><br>Presentation:<br>Delivery excellent and confident<br>Voice intonation excellent<br>Relationship with audience excellent<br>Time management excellent<br>Visual aids varied, clear and well used<br>Use of notes excellent<br>Key points emphasized and repeated |
| | 81–90 | Strong and clear structure<br>Literature extensive and fully integrated<br>Very well-developed argument<br>Original thinking/contribution<br>Data/information perceptively used<br>Very good and effective effort<br>Well written with excellent style<br>Excellent presentation | |
| | 70–80 | Strong structure<br>Very thorough use of literature<br>Very well argued<br>Originality clearly evident<br>Data/information diverse and appropriate<br>Very good and effective effort<br>Very well written<br>Very high-quality presentation | Content:<br>Strong structure<br>Very thorough use of literature<br>Very well argued<br>Originality clearly evident<br>Data/information diverse and appropriate<br>Very good and effective effort |

*Contd*

| CLASS or GRADE | MARK RANGE | DESCRIPTIVE EQUIVALENT FOR: | |
|---|---|---|---|
| | | Projects/Essays | Oral presentations |
| | | | Presentation:<br>Delivery very clear and confident<br>Voice intonation good variations<br>Relationship with audience excellent<br>Time management excellent<br>Visual aids varied, clear and well used<br>Use of notes very good<br>Key points repeated with some emphasis |
| UPPER SECOND-CLASS HONOURS | 60–69 | Clear structure<br>Good evidence of use of literature<br>Coherent argument<br>Some originality<br>Data/information diverse and appropriate<br>Good effort evident<br>Clearly and well written<br>High-quality presentation | Content:<br>Clear structure<br>Good evidence of use of literature<br>Coherent argument<br>Some originality<br>Data/information diverse and appropriate<br>Good effort evident<br><br>Presentation:<br>Delivery very clear<br>Voice intonation good<br>Relationship with audience very good<br>Time management very good<br>Visual aids varied, clear and well used<br>Use of notes very good<br>Key points repeated |

| | | | |
|---|---|---|---|
| LOWER SECOND-CLASS HONOURS | 50–59 | Some structure evident<br>Some evidence of use of literature<br>Weak argument<br>Limited originality<br>Data information incomplete<br>Limited effort demonstrated<br>Written style adequate<br>Standard presentation | Content:<br>Some structure evident<br>Some evidence of use of literature<br>Weak argument<br>Limited originality<br>Data/information incomplete<br>Limited effort demonstrated<br><br>Presentation:<br>Delivery not very confident<br>Voice intonation not varied<br>Relationship with audience limited<br>Time management could be better<br>Visual aids limited in quality and use<br>Use of notes – reads too much<br>Key points stated but not repeated |
| THIRD-CLASS HONOURS | 45–49 | Lacks a clear structure<br>Little evidence of literature<br>Shallow argument<br>No originality demonstrated<br>Data/information incomplete/ inappropriate<br>Effort level poor<br>Poorly written<br>Low-quality presentation | Content:<br>Lacks a clear structure<br>Little evidence of literature<br>Shallow argument<br>No originality demonstrated<br>Data/information incomplete/ inappropriate<br>Effort level poor<br><br>Presentation:<br>Delivery not confident<br>Voice intonation not evident<br>Relationship with audience quite poor<br>Time management needs practice<br>Visual aids limited and not clear<br>Use of notes largely read<br>Key points not clear |

*Contd*

| CLASS or GRADE | MARK RANGE | DESCRIPTIVE EQUIVALENT FOR: | |
|---|---|---|---|
| | | Projects/Essays | Oral presentations |
| PASS DEGREE | 40–44 | Not structured<br>Very limited use of literature<br>Very shallow argument<br>No originality demonstrated<br>Data/information very limited/ inappropriate<br>Little evidence of effort<br>Very poorly written<br>Poor presentation | Content:<br>Not structured<br>Very limited use of literature<br>Very shallow argument<br>No originality demonstrated<br>Data/information very limited/ inappropriate<br><br>Presentation:<br>Delivery hesitant and not clear<br>Voice intonation not evident<br>Relationship with audience poor<br>Time management poor<br>Visual aids not clear and poorly used<br>Use of notes wholly read out<br>Key points not evident |
| FAIL | 30–39 | No structure<br>Literature not used<br>Very superficial argument<br>No originality<br>Data/information inadequate/ inappropriate<br>Minimal effort<br>Very poorly written<br>Very poor presentation | Content:<br>No structure<br>Literature not used<br>Very superficial argument<br>No originality<br>Data/information inadequate/ inappropriate<br>Minimal effort |

| | | | |
|---|---|---|---|
| | | | Presentation:<br>Delivery very hesitant and unclear<br>Voice intonation monotonic<br>Relationship with audience very poor<br>Time management very poor<br>Visual aids not clear and badly used<br>Use of notes wholly read out<br>Key points not presented<br>Unacceptable presentation |
| | 20–29 | Lacking in intellectual content<br>Devoid of understanding<br>Data/information inadequate/ inappropriate<br>Fatally flawed data/information analysis<br>Little or no effort evident<br>Very poorly written and presented | Content:<br>Lacking in intellectual content<br>Devoid of understanding<br>Data/information inadequate/ inappropriate<br>Fatally flawed data/information analysis<br><br>Presentation:<br>Delivery unintelligible<br>Voice intonation monotonic<br>Relationship with audience extremely poor<br>Time management extremely poor<br>Visual aids not used<br>Use of notes wholly read out<br>Key points – none<br>Unacceptable and objectionable presentation |
| | 0–19 | Totally devoid of intellectual content<br>No understanding demonstrated<br>Data/information inadequate/ inappropriate<br>No attempt to analyse<br>Superficial and severely error prone<br>No effort evident<br>Very poorly written and presented | |

For a coursework essay that counts towards your final assessment, the marker will adopt a similar procedure, but will also make written comments on the essay for you to read. These comments are intended to help you to improve your work, and writing comments that are both honest and constructive is an important skill for a tutor. The comments should enable you to recognize how the mark you achieved relates to the assessment criteria, to see both the strengths and the weaknesses of your essay, and to understand exactly what you could do differently to improve your mark next time.

For a coursework essay that does not count towards a final assessment, sometimes referred to as a practice essay, the situation is more complex. The marker can use the essay to show you how the marking system works, how severe the standards are, and how your work measures up. That way, the marking of your 'practice' essay really is useful practice for the 'real thing'. However, unless the marking is anonymous, the marker may also use the practice essay to give you a 'lesson' mark. A lesson mark is one that is not necessarily a true reflection of what the essay deserves, but is designed to send you a message about your performance. It can be used to encourage students who need their confidence boosted, or to scare students who should be trying harder. For example, a tutor may give good students low marks as a kick up the backside to encourage them to achieve the higher standards of which the tutor knows them to be capable, and give a poor student a high mark as an encouraging pat on the back. The essay mark is sometimes used as a teaching tool.

Eric says: *The essay mark is sometimes used as a psychological weapon.*

## Is it OK to write answers that differ from your tutor's opinion?

In subjects that deal with matters of fact, and where the tutor's opinion is in line with the accepted version of 'the facts', then it would be expected that well-informed students would agree with their tutor's opinion. However, most topics in geography involve very little in the way of incontrovertible fact, and most essay questions are specifically designed to find out about the student's evaluation of a controversial issue. In most geography essays, therefore, it is not necessary that you agree with what your tutor thinks. What is important is that you explain and justify what *you* think. However, that does not mean that it's OK to think something stupid. You need to be able to defend what you think with sound evidence. Of course, you need to defend what you think with sound evidence even if you think the same as the tutor. The tutor will not be impressed just because you seem to agree. The tutor will be interested to see you explain *why* you agree. Have your own opinion, but make sure it's a good one and make sure you defend it effectively. You will not be marked down just because you disagree with the tutor, but you will be marked down if you are wrong and stupid.

Eric says: *You can think what you like as long as you're right!*

## Conclusion to Chapter 4

- Essays are marked according to clearly defined criteria. You should find out what specific criteria your tutors mark to, and you should bear these in mind when you write your essay.
- You need to convince the marker that you know your stuff, that you have worked hard, that you can organize your material to provide an effective response to the question and that you can communicate clearly.
- Now we just need to work out how we can achieve all that!

# 5 How do I choose which essay to attempt?

## *Chapter summary*

Make sure that you have properly understood what choice is available to you and make sure that you follow the rules of the assignment. In an exam, make sure you answer the right number of questions. Work to your strengths, and don't tackle questions to which you don't know the answer or topics you know nothing about. Choose titles that will give you most opportunity to demonstrate your knowledge, understanding and skills to the marker. If you can't choose, or think you can't do any of the titles, work through our checklist to help you to decide which will suit you best.

## Why do you get a choice of titles?

In an exam you are given a choice of titles so that you can make the most of your particular areas of expertise by selecting titles that suit your own interests and the topics that you have studied most carefully. In coursework there are other reasons for giving you a choice. For example, you may be using your assignment as a way of learning about a topic that you didn't previously know much about. That would not be wise in an exam, but in coursework it can make sense. So the first thing to think about when choosing which essay to attempt is your purpose in doing the assignment. Are you just out to get the best possible mark, or will other considerations affect your choice of topic? We'll come back to this in a minute, but first you need to check just what choice you really have.

## Understand what choice you really have

**The very first thing you should do when you come to choose from a list of titles is make sure that you understand exactly what options are available to you.**

Make sure you haven't already been told which essay you have to do. Make sure you really do have a choice. Read the instructions very carefully. In an exam, the instructions are likely to be set out very clearly at the top of the page, but it is surprising how many students answer three questions from section A when the instructions say quite clearly that they should answer one from section A and two from section B. Likewise with a coursework essay, make sure that you are choosing from the right list in the course handbook. Double check that your tutor has not stipulated that question 5 is only available to students in group C, or some other limitation. Check with your tutor. Check with your friends. Check, check, check. By the time you have finished this book you will be able to write a good essay. Don't throw it all away by writing the wrong one!

Eric says: *Check.*

## Know what you are trying to achieve

If you really do have a choice, you will be faced with a list of titles and you will have to pick the one you think you will be able to do best. At least, that's how most people pick their essays. You might prefer to pick the title that seems to require least work, or the one that will enable you to read up on a favourite topic. Perhaps you are writing this essay as a way of learning about a new topic. Perhaps you're practising a predicted exam question.

**When choosing which essay to do, think about what you want: an easy ride, an enjoyable task, a revision lesson, or a good mark.**

Often, these goals may coincide. For example, you are more likely to do well in coursework if you choose a topic you enjoy, because you are more likely to work hard than if you picked a miserable boring topic. Most commonly, your goal is going to be to write the best essay you can, and therefore to get the best possible mark. In our advice below, we assume that is what you are trying to achieve.

## Know your own strengths and weaknesses

Different people are good at different things. You are, no doubt, better at some things than at others. Perhaps you are good at arguing a point of logic but weak on factual knowledge. Perhaps you have a good grasp of grammar but have trouble keeping yourself focused on one point at a time. Perhaps you know a lot about urban regeneration but nothing about deserts. That's OK. At this stage what matters is that you know your strengths and weaknesses so that you can choose titles that give you the opportunity to use your strengths and avoid titles that hit your weak side. If you don't feel confident that you know your strengths and weaknesses, look at the feedback that you got from previous pieces of work. One of the tutor's jobs in assessing your work is to tell you what you have done well and done badly. This isn't just for the tutor's amusement, it's to help you to recognize your strengths and weaknesses. In the long run it would be a good idea to work on correcting some of your weaknesses, but if you are facing a choice of essay titles right now it's a little late to start a personal development programme. Acknowledge your weaknesses and avoid them.

## Work out which essays you should not attempt

In any list of titles there will usually be some that would get you a good mark and some that wouldn't. Your problem is to figure out which are which. Step one is to eliminate the titles that you should avoid. It's like crossing off the no-hopers in a list of runners for a horse race. There are two simple rules that will help you to avoid titles that will get you into trouble (Box 5.1). These rules are so simple that you might wonder why we have to explain them, but we have seen hundreds of students come unstuck because they ignored this common-sense advice.

**Box 5.1 Two fundamental rules for identifying titles to avoid**

- Don't tackle questions to which you do not know the answer or titles where you cannot do what the title instructs (e.g. discuss, explain, etc.)
- Don't tackle topics you know nothing about.

### Don't tackle questions to which you don't know the answer or titles where you cannot do what the title instructs

The title of your essay requires you to do something: it is either a question that demands an answer (what? how? when?) or an instruction that needs to be fulfilled (explain, discuss, evaluate). If you can't actually do what the title demands, then you must not choose that title. Even if the topic is one that you are familiar with, even if it's the topic you have revised especially, if you can't do the specific thing that the title demands, you must move on down the list. Here's an example of what we mean. Suppose you have learned all about glacial moraines and revised them for the exam. Joy of joys, there is a question on moraines on the paper. The question says: 'Compare the internal structure of lateral and terminal moraines.' The problem is, you didn't know that there *was* a difference between the internal structures of those types of moraine. It's your topic, and you have memorized a load of information about it, but you can't answer that specific question. Don't choose that essay. The marker won't be impressed by the fact that you know other things about moraines. Once it becomes clear that you don't know the answer to this specific question, you're sunk.

## Don't tackle topics you know nothing about

Needless to say, if you don't know anything about glacial moraines, don't waste too much time in an exam wondering whether to do 'Compare the internal structure of lateral and terminal moraines'. Obvious, perhaps, but not to everyone. It can be tempting sometimes, if you just happen to think you could maybe figure out the answer to a question, to have a go on the spot without any background knowledge. Sadly, even if you know the answer to the specific question, but have no factual background or deeper appreciation of the topic, the marker is unlikely to give you more than about 20% just for getting the answer 'right'. Most of the marks a tutor can give are for setting your answer into its context: knowing something about the literature, seeing how the question is relevant to other issues within the discipline, relating your case study location to contrasting sites that provide a different viewpoint on the question, and so on. If you know nothing about the topic other than the basic answer to the question, you will miss out on all those marks. In a coursework essay, of course, this rule may not always apply. The essay may be set before you have had many lectures or done much reading, so it is inevitable that you won't know much about the topic. The trick in this situation is to remedy your ignorance: find something out about the topic before you start writing an essay on it.

Eric says: *If you don't know anything about population, don't tackle questions about demographics.*

# Work out which essay you can do best

By now you may have struck a few titles off your list of options. (If you have struck off all the titles, don't panic – go to the 'Emergency scenario' section later in this chapter and we'll sort something out for you.) Your next job is to figure out which of the remaining titles you could do *best*. Remember, what you are trying to do here is convince the examiner to answer 'yes' to all those questions that examiners ask themselves, which we covered in Chapters 3 and 4 (Box 3.2). You want to use your essay as a vehicle to convince the marker that you have knowledge, understanding and skill; that you can see clearly the relationships between broad issues and specific details; that you have done some research, given the subject some thought, and concocted something worthwhile to say on the topic. You can use many of the items from Box 4.1 as a checklist to compare your ability to deal with each of the remaining titles on your list. We have repeated some of the key points here, and added a few more to help you to decide whether each essay on your list is 'the best' for you (Box 5.2). For each title, check how many of these questions you can answer 'yes' to.

**Box 5.2 For the essay you are thinking of choosing**

- Do you understand the question or instruction?
- Can you identify the key issues raised by the title?
- Can you answer the question or fulfil the instruction in the title?
- Can you summarize your answer in one sentence?
- Can you place your answer into the broader framework of the subject?
- Do you remember the course material on which the title is based?
- Have you worked independently beyond basic course material on this topic?
- Can you put together a logical and coherent argument to make your point?
- Can you provide evidence to support your point?
- Can you correctly cite the sources of your evidence?
- Are you widely read and well informed?
- Can you structure your argument and your evidence into a coherent essay?
- Can you think of any diagrams, quotations or other illustrations to use?

Ideally, you should not start writing an essay unless you can answer 'yes' to all of these questions. If you cannot answer yes to the first three questions, this title should not have made it through the first elimination round. Cross it off your list. The fourth question in Box 5.2, 'Can you summarize your answer in one sentence?', is designed to check that you are not just *saying* you can answer the question without really thinking about it. If you really do know the answer to a question you should be able to summarize that answer in one sentence. If you think you know the answer, do the sentence.

Eric says: *Work your way down your list of possible titles, trying to write the one-sentence answer to each one. If you can't do the sentence, don't do the essay!*

When you have checked each possible title against this list (or a similar list of criteria that you could make up for yourself), you will probably be left with a shortlist of titles that score full marks. These are the serious candidates for your attention. If there is only one title on your shortlist, then your choice is simple. If there are several, it may at least be clear that your 'yes' answers were more enthusiastic for one title than for the others, and you can decide on that basis. If there are several titles for which you are equally enthusiastic, you need a tie-breaker to sort out the winner. Surprisingly, perhaps, the best advice in this situation is to choose the most 'difficult' title.

'Difficult' is an odd word to use since we've just established that you are equally enthusiastic about each of the titles left on your shortlist. What we are getting at is that some titles are clearly complex, pointing to a lot of not-so-hidden depths in the topic, while other titles are superficially much more straightforward. We will discuss different types of title in Chapter 6. For now, suffice to say that 'easy' questions make it difficult to excel, difficult to dazzle the marker with your virtuosity. Even students who are less able than you will be able to have a good stab at 'List the landforms associated with permafrost'.

**It will be hard to seem brilliant in your response to an easy question.**

By contrast, you will have much more opportunity to impress the examiner with logic, depth of insight, and references to recent research discoveries if you tackle: 'Assess the relative merits of geochemical and sedimentological approaches to differentiating ground ice from buried glacier ice in permafrost environments.' More complicated questions are sometimes easier to score high marks on. They give you more to get your teeth into. They give the marker more opportunity to be impressed. If you've got to the stage of having several competing titles on your shortlist, then you're good enough to tackle the tough ones.

*Eric says:* *Be adventurous. Go for it!*

## Emergency scenario: what if I can't do any of them?

Let's suppose you work your way through the list of options and you eliminate all the titles. You don't think you can do any of them. In a coursework essay, this situation should not be a real problem, because you can go to the library, find out more about the topic, and get yourself into a position where you *can* do one of them. All you need to do is work out why you can't do it, and then tackle that obstacle. If you are limited by factual ignorance, look stuff up. If you are limited by blank incomprehension when staring at the question, talk to somebody about it. Start with a friend, and if necessary end up going to see the tutor.

**In an exam situation, usually made worse by nerves and pressure, thinking that you can't do any of the essays on the list is a very common experience.**

This is such a big issue that we have a whole chapter on exam essays for you to look at later in the book (Chapter 16). The first important thing is to stick to the rules and make a choice according to the rubric of the exam. If you are supposed to do three questions but can only find two that you like, you must do a third one anyway. Do not believe that you will make up enough extra marks spending extra time on your two good questions to make up for missing out that third one. It doesn't work. After working on an exam essay for an hour, you've probably got all the marks you are going to get. Keeping at it for another 10 minutes won't raise your score very much. Starting your next essay, even if it's a weak one, you will probably earn several grades in the first 10 minutes of writing.

If you are completely stuck, even after you calm down and have a second look, you will just have to lower your 'acceptance criteria' and re-apply your list from Box 5.2 in a less stringent fashion. Box 5.3 suggests a possible emergency test that you can apply to the full list of titles once you discover that none of them gets through the standard shortlisting procedure.

### Box 5.3 Minimum requirements when you're really stuck

- Do you understand the question or instruction?
- Can you identify some issues relevant to the title?
- Can you place those issues into the broader framework of the subject?
- Do you have any factual knowledge at all to support your point?
- Can you offer any answer to the specific question or instruction in the title?

If you can answer 'yes' to just these questions, then you can have a go at the title and be confident that you should at least raise some kind of marks. It isn't going to be pretty, but it's better than nothing.

If you can't find a title to which you can answer 'yes' to all these questions, then things are going to get a bit sticky. If it's coursework, you're OK because you can go and do some reading, have a good long think, and ask for help. As long as you haven't left it to the last minute you should be OK. Just pick the simplest-looking title and keep working on background material until you can answer 'yes' to the questions. If you *have* left it to the last minute, it's late at night and your essay is due in tomorrow morning, and you can't answer 'yes' to the questions, then you are in serious trouble. You need to refer immediately to the Disaster Kit in Chapter 21.

If you're in an exam and you can't find a question that even the reduced checklist will allow you to do, then it's pretty much too late for you and there's nothing much that we can do. This is a situation that you really need to avoid. If you've read this book and remembered the emergency advice that we give at key points (especially in Chapter 16, about exams, and in Chapter 21), then you might be able to rescue something, but you really shouldn't get yourself into that position. The owners' manual for a German sports car that one of us owned explained that drivers should take care going round corners because although the tyres, suspension and braking systems were of superior technological design, technology could not overcome the physical limits of roadholding. In the same way we should warn you that although our advice may help you to improve your performance, it cannot overcome the physical limits of essay writing. If you go into an exam wholly unprepared and woefully ignorant you shouldn't be surprised that you can't answer any of the questions, and we will not be able to weave any magic to save you. Think ahead.

Eric says: *Don't say they didn't warn you!*

## Conclusion to Chapter 5: 5-minute procedure for choosing which question to do

- Read, check and double check the instructions about choice of title.
- Eliminate any questions that the instructions don't allow you to attempt.
- Eliminate questions on topics you have not studied.
- Eliminate questions to which you do not know the precise answer.
- Eliminate questions for which you do not have case studies and references.
- Write one-sentence answers to the remaining questions.
- Pick the most interesting sentence and write that essay.
- In case of a dead heat, choose the most challenging title.

# 6 How do I get started on my essay?

## Chapter summary

A coursework essay is a lengthy project, and eventual success depends on good groundwork in the early stages. Getting started on your essay involves two main steps: getting to grips with the question and getting to grips with your answer. As soon as the assignment is set you should take time to understand exactly what the question requires you to do. You should then let it simmer in the back of your mind for a while before starting to work on gathering source material and planning your answer. The instructions given in an essay title are carefully set and must not be ignored. They will tell you what type of essay you have to write. Start by making absolutely sure that you are doing what you are supposed to.

## Timing: when to do what

Before thinking about *how* to get started, it's worth deciding first of all *when* to get started. Coursework is typically set a long time in advance of the submission deadline. You may have a few weeks to put together an essay, or even a few months. The reason that you are given so long is because putting together an essay is a long job. We don't just mean that you need to find time in your busy schedule for 10 hours of reading and 10 hours of writing. We mean that you need extended periods of time to turn material over in the back of your mind and slowly develop your ideas about the question. If you don't start to think about your essay until a couple of days before it is due in, your answer is likely to be less well developed than the answers of students who started thinking about it weeks earlier. Those students may not have finally sat down to write the essay any earlier than you, but their minds have been working on it for weeks while yours has only been working on it for hours. You don't have to put in more hours of slog, but you need to put in your hours of slog at the right time. In this chapter we'll go through the early stages of the process.

Box 6.1 How to get started on your essay

*There are five steps to getting started on an essay.*

1. Memorize the title.
2. Get to grips with the question.
3. Understand what the question is really asking you to do.
4. Do nothing for a while.
5. Get to grips with the answer.

It is important to get these in the right order.

## The first job: memorize the title

Once you have decided, or been told, the title of your essay, the first thing you must do is commit it to memory. This may sound silly, as you already have it written down in your handbook or in your notes, but it really is important. We want you to keep the title in the back of your mind so that it pops up out of your subconscious whenever a relevant bit of information jumps out of a lecture or a book or a conversation. For this to work properly you need to have the question firmly embedded in your memory. It's a good idea to begin by writing the question on a bit of paper that you keep in your pocket. Write it on your fridge door, paint it on your bedroom ceiling, tattoo it on your best friend's face.

**When you copy down your title, make sure you copy it down correctly.**

It is amazing how many students hand in essays with the title copied out slightly wrong. If you answer a question slightly different from the one that was set, your answer will be slightly different from the one the tutor expects. Tutors go to a lot of trouble to word their questions very carefully to say precisely what they mean. Altering a word, or a punctuation mark, can make a big difference. There is no point letting your mind get to work on the wrong question. Check out the examples in Box 6.2 to see how much difference a small error in copying can make. Be careful.

## The second job: getting to grips with the question

One of the most common causes of failure in essay writing is misunderstanding the question. In extreme cases this can involve a complete misreading, where the student writes about something quite different from what the tutor intended. More commonly it involves one of two things:

1. The student fails to spot some subtlety in the phrasing of the question.
2. The student fails to recognize the part of the topic that the question is really getting at.

## Box 6.2 Examples of small errors in copying

*Small errors in copying leading to big changes in the meaning of a question.*

*Example 1:*

Is 'health for all' an achievable goal for urban areas in the developed world?
Is 'health for all' an achievable goal for urban areas in the developing world?
Is 'health for all' an achievable goal in the developing world?

*Example 2:*

Examine the processes controlling landforms in cold arid environments.
Examine the processes controlling landforms in cold or arid environments.
Examine the processes controlling landforms in one cold arid environment.

You need to think very carefully right at the outset about what the question really says, and what the question really means. There are a number of things to think about.

First, there are many different types of question, and different types of question require different types of answer. You need to identify what type of question you are dealing with. Second, some questions have quite complex structures, and you need to decipher the question to make sure you have understood how all its parts fit together. Third, when you've recognized the type of question you are dealing with and deciphered the structure of the question, you need to make sure that you understand exactly what the question is asking you to do. Then it is safe to proceed.

Before you do anything else, make sure that you really understand exactly what the question is asking you to do. Remember that the important words in the title are the words like 'discuss', 'describe' and 'explain', not the words like 'glacier', 'urban' and 'banana'.

## Different types of question

Different types of question require different types of answer. If the question is 'Why . . .?' then the answer must be 'Because . . .'. If the question is 'Does . . .?' then the answer must be something like 'yes', 'no', 'sometimes', or 'it depends'. If the question says 'Compare A and B' then your answer should compare them. If the question says 'Evaluate the relative importance of X and Y' then the only worthwhile answer is one that evaluates their relative importance. If the question says 'Why is A bigger than B?' and you answer 'It is very significant that A is much bigger than B', you will probably score close to zero.

**You need to recognize the type of question you are dealing with, and provide the appropriate type of answer.**

Usually, there is a very clear key word in the title that indicates what type of question is being asked. Explain. Describe. Evaluate. Discuss. Sometimes different key words indicate similar questions. For example, 'Why is A bigger than B?' is effectively the same question as 'Explain the difference in size between A and B.' In both cases the answer will be of the form 'A is bigger than B

## Box 6.3 Different questions on the same topic, and appropriately different styles of answer

Q: Why do landforms change through time?

A: Landforms change through time because . . .

Q: Evaluate the assertion that all landforms change through time.

A: This assertion is (valid / invalid / valid in some ways). The assertion applies to timescale *x* but not to timescale *y* . . . The assertion would be better stated as . . .

Q: Describe how landforms change through time.

A: Landforms change through time in these ways, at these rates, on these spatial and temporal scales . . .

Q: 'Landforms change through time.' Discuss.

A: This is generally true, but the details are controversial . . . Some research has shown *x*, but research elsewhere has shown *y*. The scientific implication of this controversy is that . . .

## Box 6.4 Examples of inappropriate answers to different essay questions on the same topic

*Compare these answers with those in Box 6.3 and put yourself in the place of the tutor marking these answers.*

Q: Why do landforms change through time?

A: This is generally true, but the details are controversial . . . Some research has shown *x*, but research elsewhere has shown *y*. The scientific implication of this controversy is that . . .

Q: Evaluate the assertion that all landforms change through time.

A: Landforms change through time in these ways . . .

Q: Describe how landforms change through time.

A: The assertion that landforms change through time is valid . . .

Q: 'Landforms change through time.' Discuss.

A: Landforms change through time because of *x* and *y*.

*because . . .*'. Box 6.3 gives a list of essay titles of different types, along with the type of answer that each demands. Work your way through the list and make sure that you understand how the answers match the questions. Afterwards, look at Box 6.4 and imagine that you are a tutor giving marks to the answers to each essay in that list.

## What's the difference between 'describe', 'discuss', etc.?

Very careful thought will (or should) have gone into devising the exact wording of your essay title. Consider the essay titles in Box 6.5. Thinking about the appropriate and inappropriate styles of answer we looked at in Boxes 6.3 and 6.4, would you write the same answer to all of these essays?

### Box 6.5 Different essay titles on the same subject

- Describe the processes of soil erosion.
- Illustrate the processes of soil erosion.
- Discuss the processes of soil erosion.
- Evaluate the processes of soil erosion.

We hope you replied 'no' to the previous question. Although each title is about soil erosion, none of them says 'Write all you know about soil erosion'. Instead each of them gives you a specific instruction to write one aspect of what you know. The words 'describe', 'discuss' etc. are signposts to the type of essay your tutor is looking for. They tell you the questions that you need to ask yourself if you are trying to decide whether to answer this question or not, or how you should go about answering it if you have no choice.

'*Describe*' is easiest. The only questions you need to be able to say 'yes' to before you can tackle this question are 'Do I know what the processes of soil erosion are?' and 'Do I know enough about them to fill an essay of the appropriate length?'.

'*Illustrate*' is asking for a bit more. Not only do you need to know what the processes of soil erosion are, but to answer this one you also need to have some good case studies that you can refer to in some detail. Of course, having case studies will also help with the 'describe' question, but they are not obligatory. With 'illustrate', if you don't have the case studies you can't do the essay.

'*Evaluate*' is asking for some quantitative comparison of the different processes. You will be expected to show that process $x$ leads to $n$ times as much soil loss as process $y$, but that process $y$ affects $m$ times more of the world's soil surface than process $x$, for example. So now you have to be able to answer the questions

'Do I know what the processes of soil erosion are?', 'Do I have estimates of how much soil loss they each cause?' and 'Do I know how much area is affected by each process?'.

'*Discuss*' is a different kettle of fish. This isn't just a signpost; it's a flashing electronic display board. It tells you there is controversy, that different people have different opinions, that there are conflicting data. All of these things will need to be covered in your essay if it is to be good. To answer the 'describe' question you could probably get away with going to one textbook on soil erosion where you would find enough to produce a competent answer. For the 'discuss' essay one source of material will certainly be insufficient.

**For '*Discuss . . .*' you will need to have read widely, and to have ferreted out the controversial views.**

However, you don't have to decide who is right and who is wrong. If the experts don't agree, it's because there are significant opportunities for different interpretations of the available information. To satisfy the instruction 'Discuss . . .', it is sufficient for you to report the details of the controversy. If you want to make a judgement based on what you have read, by all means do so. But make sure it's a judgement that you can back up with a reasoned argument, and not just a preference for the views of the person born in the same town as you.

In Box 6.6 there's a list of the most common instructions you are likely to see in essay titles and our interpretation of what they require you to do.

## Box 6.6 Instructions given in essay titles and what they mean

| | |
|---|---|
| *Describe* | Tell us what it's like. |
| *Explain* | Tell us why. |
| *Discuss* | Tell us about different ways of looking at it. |
| *Evaluate* | Tell us whether it's any good, i.e. what value it has. |
| Assess | Pretty much the same as evaluate. |
| *Illustrate* | Give us some good examples of it. |
| *Review* | List, describe and assess. |
| *Comment* | Explain, discuss and evaluate. |
| *Examine* | Describe, explain and discuss. |
| *Write about* | Describe, explain, discuss, evaluate and illustrate. |
| *Compare and contrast* | Show what's similar and what's different. |
| *Critically discuss* | Discuss and evaluate. |
| *Account for* | Explain. |
| *Outline* | Describe the key features of. |

NB: Assume that all titles include the phrase 'with reference to specific examples'.

## Why am I sometimes asked to describe something and at other times to discuss it?

Whether you are asked to describe something or discuss it will depend on which of the things we listed in Box 3.2 are at the forefront of your tutor's mind in setting you the essay. If the key things are assessing your knowledge of the subject and your ability to follow conventions in writing in your discipline, then you are likely to be set essay titles of the 'describe' type. If the key things are assessing your ability to use evidence to argue a point, or determining how widely read you are, then essays of the 'discuss' type are more likely. You can also expect to see a progression as you go through your course. In your first year, the emphasis is more likely to be on getting you to be able to write essays well and communicate clearly. For those purposes, there's no need to clutter up the essay with a complex and difficult instruction. But towards the end of your course in your final year, your tutors will be looking for the sharp, incisive trained mind that you should, by then, have developed. The essay titles you will be given at that stage will be designed to sort the sharp sheep from the gormless goats.

Eric says: *Try to be a sharp sheep, not a gormless goat! Be wheat, not chaff.*

## Deciphering complex question structures

Some questions are constructed in a very simple style. 'Why do some glaciers move faster than others?' 'Describe the El Niño phenomenon.' 'Discuss the impact of the motorway network on rural traffic density.'

Other questions are formulated in more complex fashion. For example: '"When James (1995) wrote of the 'urban conscience' he neglected the impact of the suburban" (A. Smith, 1999). Evaluate Smith's critique with specific reference to James's original case study location, and consider whether it applies equally to cities in the United Kingdom.'

When faced with a complex question like that, the need to work out exactly what you are being asked to do is very apparent. You should get into the habit of thinking about what the question is really asking even when you look at 'simple' titles. It's always possible that the focus of the question is not really on the most prominent subject in the title. For example, in the 'urban conscience' question, to what extent are you being asked to comment on James's original work? What does the word 'it' apply to in the last part of the question? In the rural traffic density question, to what extent are you being asked about the motorway network?

Eric says: *Work away at the question with a needle to separate the meat from the bone.*

## The third job: understanding what the question is really asking you to do

Once you've spent some time dissecting the question to identify the key words, separating the core of the question from the skeleton of the supporting words, you should be in a position to know what the question is really asking you to do. Is it asking you to explain something, or describe something, or compare two things or three things? Is it asking you about James (1995) or Smith (1999)? Did you work out what the 'it' referred to? Should you be focusing on the motorway network or on rural traffic density, or both? For your particular title, you need to work out all these types of queries.

Now for the first time you are ready to put pen to paper in earnest. You should write out your question (again) and immediately below it write an example of what your one-sentence answer could be. At this stage we don't expect you to be able to write the *actual* answer, because you haven't done any real thinking or research yet, but you should be able to recognize what the answer could, and could not, look like. For example, the urban conscience title is one that we dreamed up out of thin air, so there is no real answer (there is no real James or Smith), but from the structure of the question we know what the answer must look like. The answer must take the form: 'Smith's critique is valid/not valid for James's original case study location, and/but does/does not apply equally to UK cities.' There may be more to it than that, and there will be discussion and elaboration, but that must be the basic form of the answer. There are other answers we could write that would not be appropriate. For example: 'James's idea of "urban conscience" was limited by his preoccupation with the rural idyll, and has never been applied to cities outside Canada.' This answer focuses on James when it should focus on Smith's critique, and it doesn't discuss whether Smith's critique would be valid for UK cities. The question didn't want you just to write about James, Smith and their ideas. It specifically wanted you to answer a very precise question.

**If you answer the precise question that was asked you will be OK. If you don't, you won't.**

For the essay that you are working on, write down the question and your proposed type of answer on a piece of paper. Check that the answer answers the question. Take the bit of paper to show your friends. What do they think? Take it to show your tutor. If you get this bit wrong you will inevitably do badly. If you get it right, you have made a good start. Get to that stage before you proceed.

## The fourth job: after you've made a start, do nothing for a while

Most students think that doing nothing is step one! It isn't. Once you have worked out what your essay is about and memorized the title, then you can *usefully* do nothing for a while.

Eric says: *Can I say in my tutorial report that I've spent the week usefully doing nothing?*

The idea of doing nothing for a while is to let the essay simmer away on the back burner of your subconscious. During this time, everything that you read, or hear in a lecture, or see on the television news, or notice in the world around you will be filtered through the subconscious of the same mind that is subconsciously remembering your essay. For most people, this has two effects. First, it means that when you turn your attention back to your essay a couple of weeks later you will miraculously discover that the title makes much more sense than it did when you first saw it and that your mind is full of bright ideas about how to answer it. Second, it means that whenever something relevant to your essay turns up in lectures, on TV, or in the pub, you will notice it. This puts you at a huge advantage over your colleagues who have not remembered their essay title and therefore do not notice the relevant information when it passes them by. During this period of 'doing nothing' on your essay, you will no doubt be working on other things, but it wouldn't hurt to keep a page of notes running for ideas that strike you about your essay. Once things start simmering away in your subconscious, ideas will pop up from time to time and if you note them down in your file you will be able to remind yourself of them later, when you settle down to get to grips with putting together your answer.

## The fifth job: getting to grips with the answer

If you have followed our advice you will by this stage have identified your question, worked out exactly what it means, recognized the style of answer that you have to produce, and let it simmer in the back of your mind for a little while. During that time you may have had a few ideas, and you may have noticed some relevant source material in lectures, in the literature, or in instances from your daily experience. This is a good starting point for getting to grips with your answer. This is the second stage of 'getting started' on your essay.

Getting to grips with your answer involves several steps. In the next chapter we will talk about setting up the structure of your essay, devising an essay plan, and using the plan to build a sound essay. However, before we can do that we need to decide what the answer actually is.

We already know what *type* of answer you need to produce, because we thought about that when we identified what type of question we were dealing with, but as yet we have not established what the *actual* answer is. We may know the answer is 'Because . . .', but we don't know what the '. . .' is!

Eric says: *This is the difficult bit, isn't it?*

Arriving at 'the answer' involves a combination of knowledge about the topic and reasoning about that knowledge. The reasoning bit is very important, but you must have some information to reason about. You can't explain why some glaciers move faster than others if you just don't know what controls the speed at which glaciers move. You can't convince the examiner that you have a worthwhile interpretation of Smith's critique of James's theory if you don't know what Smith's critique was. You can't convince the examiner that you are an expert on El Niño if you have not read different people's explanations of the phenomenon.

**Basically, now is the time that you have to know your stuff.**

Information that you need to provide ammunition for your essay can come from many different sources. Basic material will probably have been provided in lectures and more will come from the associated recommended reading. Further material will come from your independent research beyond the core recommended sources. Identifying all the possible sources of geographical information is beyond the scope of this book, but probably at this stage you need to go away and do lots of reading.

Most essays require you to do more than simply report information that you have derived from lectures and from your reading. Usually you are expected to use that information as a means to an end, rather than as an end in itself. The end that you are heading towards is usually a reasoned argument about the issues raised by the title of your essay. Referring back to our 'urban conscience' example, knowing about James's theory and Smith's critique is essential, but you then have to go beyond that knowledge to give your own argument about the theory, the critique, and their wider applicability. So, the 'raw materials' that you need to go away and collect before we go on to Chapter 7 include both information about the topic and your own thoughts about that information in so far as it relates to your title. Go away and collect all the ingredients listed in Box 6.7.

## Box 6.7 Materials to gather before you start to construct your essay

- Facts and figures and names and dates
- Other people's opinions
- What you heard in the lecture
- What you read in the recommended reading
- What you read in your independent research
- Major disagreements between sources
- Good examples
- Good diagrams
- Topical case studies from current affairs
- References from the most recent top-flight journals
- References from the all-time classic sources
- Your thoughts about each of the above
- Your friends' thoughts on each of the above.

## *Conclusion to Chapter 6*

- Before you start to construct your essay, you must establish a good foundation.
- First, understand the question. Second, gather all the raw materials that you will need to plan your answer. The most important raw materials are clear thinking and a good knowledge base.
- It is as important for you to understand the specific instructions given in the title of your essay (describe, discuss, etc.) as it is to know the topic of the essay. Use the instructions as clues to the type of essay your tutor is looking for.
- If you are starting work on an essay now, make sure you have reached this stage of preparedness before you launch into writing.
- When you are ready to start, move on to Chapter 7.

# 7

# How do I make sure I have a well-structured essay?

## Chapter summary

**Designing a good structure is about making sure that all the components of your essay are put together properly in the right order. This includes both logical components (such as questions, evidence and conclusions) and physical components (such as paragraphs, references and illustrations). The basic structural components of most essays are the introduction, the middle bit and the conclusion. However, specific types of answer such as the list, the explanation and the comparison have specific structural requirements. The production of an essay plan is a key stage in the design of a good structure, and the separation of the text into sections is the fundamental mechanism by which you give physical structure to your essay. Your essay plan should be sufficiently comprehensive that you know exactly what your essay will look like before you start writing it.**

## Do I need to read this before starting the essay?

Essay 'structure' is one of the things that people always go on about when they are telling you how to write essays and when they're telling you what you did wrong when you wrote your last one. It's one of the main criteria in most formal marking schemes. It's one of the things that the person marking your essay will have to pass judgement on, and it will probably be one of the things that somebody will be going on about when you get feedback from your next essay.

**Structure is clearly an important topic, but it's one that many students never have satisfactorily explained to them.**

When building something complicated it's important to have a clear idea of the overall structure before you start putting bricks on top of each other. In your essay you could think of each sentence as a brick, so before you start putting sentences in sequence one after the other you need to think about the overall structure of the essay. In this chapter we will explain what 'structure' is, and show how you can make sure that your essay has a good one before you start writing. So, yes, you should read this before you start!

# What is essay structure?

An essay is made up of lots and lots of little bits. There are lots of separate little ideas, lots of bits of evidence, a bunch of examples, a handful of case studies, three or four major themes, an answer, a contrasting viewpoint, an introduction, a couple of dozen references, perhaps a hundred sentences, a few diagrams, maybe 1500–2000 individual words, and many other components. It's like the list of ingredients for a recipe, or a load of materials at a builder's yard. You could put them together in any order. It's pretty obvious that if you jumble up the words into the wrong order the essay is not going to make much sense. However, it isn't always as clear to see whether the essay will still make sense if you put the three or four major themes in a different order. And what if you put the contrasting viewpoint at the start of the conclusion instead of at the end of the introduction? What about swapping paragraphs 5 and 6? What about putting the answer right at the start of the essay instead of saving it up for the conclusion? The order of the parts does make a difference.

**You need to put everything in the right order.**

The list above covers both physical components of the essay (words, sentences, paragraphs) and logical components (answer, evidence, counter-argument). It is important that both the physical and the logical components are in the right order. We will worry more about the physical components in Chapter 8 when we think about paragraphs, sentences, grammar and syntax. Before you start worrying about your use of words, you need to get the structure of your *ideas* right. The most fundamental aspect of essay structure is the logical sequence of your argument. Two main things can go wrong here:

1. The components of your argument can be presented in an unhelpful order. Look at Box 7.1 for an example.
2. The argument can be constructed illogically so that your answer does not follow reasonably from the argument you present. Look at Box 7.2 for an example.

Your essay needs to have a sound, logical argument leading up to your answer, and it needs to be put together with all the parts of the argument in the right order. It is also important that the physical components of the essay (words, sentences, paragraphs) are in the right order, and that each part of your logical argument is supported by an appropriate set of physical components. For example, if your first main theme is packaged in a set of three paragraphs each made up of several sentences, then your second main theme should not be squeezed into a single sentence at the end of one of those paragraphs.

## Box 7.1 What happens if you put the components of your answer in the wrong order?

If someone asks you what time it is, your reply may be a simple statement: '3 o'clock'. This does not take much organizing. However, if you don't know the time but you know the whereabouts of a clock, you need a more complicated response consisting of several parts. In many essays the student seems to know the relevant information but just puts it in an illogical sequence and misses bits out.

Q: 'Excuse me, could you tell me the time?'

A: 'There is a shoe shop 50 m down this road. There is scaffolding on the front of the building. I don't know the time. The scaffolding doesn't cover the clock.'

This answer could be deciphered (by an experienced essay marker) but is less helpful than it should be simply because the structure is wrong. The answer works best if you put the parts in a logical sequence.

Q: 'Excuse me, could you tell me the time?'

A: 'I don't know, but there is a clock outside the shoe shop. If you go there you will be able to see the time. The shop is 50 m down this road, and the clock is just above the scaffolding that covers the shop window.'

## Box 7.2 What happens when your argument is illogical?

Q: What kind of animal is Fido?

A: Fido is grey. Elephants are always grey. Therefore Fido must be an elephant.

Q: What is this ridge of sand in the Sahara Desert?

A: This ridge is made of sand. Eskers are made of sand. Therefore this ridge must be an esker.

From what we've said so far, structure seems to be about:

- putting things in the right order;
- being logical in your explanations;
- making sure that all the components of your essay fit together properly.

We don't want the foundations on top of the roof, and we don't want one side of our square to be longer than the other three. In theory, this all sounds quite reasonable. How do we apply that to specific essays?

## Should my essay have an introduction, middle and conclusion?

Essays traditionally have an introduction, a middle bit and a conclusion. This time-honoured structure is generally sound and provides a good base for developing different essays. We will stick with it, but you should remember that each part can be quite different from essay to essay. Over the next few chapters we'll take each part in turn, but to begin with let's just agree that in some form or other our essay will have an introduction at the beginning, a middle section in the middle, and a conclusion of some kind at the end. That's the very first step in deciding our essay structure. It's so fundamental we should put it in a box and keep it (Box 7.3).

### Box 7.3 Essay structure for absolute beginners

1. Introduction
2. Middle
3. Conclusion.

Having done that, we can forget about the introduction and conclusion for a while. Our first job will be to design the structure of the middle bit, in other words the organization of our core material. Once we have worked that out, the structure and content of the introduction and conclusion will follow logically from what we have planned to put in the middle.

## Do different essays require different types of structure?

We established in Chapter 6 that there are different types of question, and that different types of question require different types of answer. These different types of answer require different structures.

Different types of structure include the list, the explanation, the comparison, the assessment or evaluation, and the discussion or analysis. Box 7.4 gives some examples of the kinds of question that could be tackled with each kind of structure, and gives an indication of what the answer would look like in each case.

## Box 7.4 Different essay structures, what they look like, and the types of question they can be used for

**The list**
*Structure:* A, B, C, D and E
*Example question:* 'Identify the main elements of . . .'; 'List the causes of . . .'

**The explanation**
*Structure*: X is Y because A generates B, and consequently C causes D.
*Example question*: 'What causes . . .?'; 'Why . . .?'; 'Explain . . .'

**The comparison**
*Structure*: A is like X, but B is like Y.
*Example question*: 'Compare . . .'

**The assessment or evaluation**
*Structure*: A is good for both X and Y but is no use for Z.
*Example question*: 'Evaluate . . .'

**The discussion or analysis**
*Structure*: A depends on X and causes Y.
*Example question*: 'Discuss . . .'

**The ramble (this is a bad thing)**
*Structure*: B, A, D leads to G. R is like A. D, E and S. Also X, no, X.
*Example question*: can be applied, unfortunately, to any question.

Some of these structures are horizontal, or in series, which means that the components run along one after the other like carriages in a train. The list, for example, runs along in series and it doesn't even matter, sometimes, what order you put the carriages in. Some simple explanations can also run as a horizontal series, but for explanation you normally have to put your points in a particular order for them to make sense.

Other structures are vertical, or in parallel, which means that different parts of the structure need to be brought together at some place in the essay in order to make a point. The assessment type of question, and most types of explanation, are like that. You may need to use the fact that A is good for X and Y to explain why it is no use for Z. In that situation, what you write about Z depends on what you have already written about A, X and Y. You are building a tower where the pinnacle depends on the foundation.

Some of the structures in Box 7.4 involve both series and parallel elements. For example, most discussion or analysis structures involve both elements of listing and elements of explanation. In these essays particularly, it is important to keep the structure clearly defined. If you don't, you could find yourself producing a 'ramble', which is what happens when the structure collapses and parts of the answer are wheeled in on stretchers and dumped in the emergency room of your crumbling essay.

# How do I decide on a structure for my particular essay?

If the basic structure of your answer is determined largely by the style of the question, as we suggested in the previous section, you should be able to decide what type of structure your essay will adopt as soon as you get to grips with the question (see Chapter 6). If your question calls for a list-like answer, then you need to employ a list structure with carriages lined up one behind the other. If your question demands that you explain something complicated, then a vertical explanation structure is called for, with the tower of the argument building up level by level on strong logical foundations. If you have worked your way through the stages that we have suggested of getting to grips with the question, getting to grips with the answer, and then thinking about overall structure, it should not be difficult to decide on an appropriate broad structure for your essay.

In fact, when you have had some practice writing essays you will learn which kinds of structure suit you best and this knowledge will help you to select which title you might choose for your assignment.

**Some students excel at complex discussion and analysis, while others are more comfortable with the humble list.**

By the time you get to an exam you need to have found out if you have particular weaknesses in certain types of essay. That way you can decide on the kind of essay structure you want to work with, and then find a title that will enable you to do so. The difficult bit starts when you try to convert that idea of a broad structure into a more specific plan.

# What if the style of the question does not suggest a structure?

Sometimes, you may be set an essay title that does not obviously lend itself to any particular structure. For example, 'Write an essay on soil moisture' would be one of these questions. 'That's easy, I can't go wrong', you might think. But ask yourself why your tutor would set such an essay. Actually, your tutor is being even meaner than normal by withholding the usual clues that titles often give about how to approach the essay. You are being asked to do all the things listed in Box 3.2 *and* to work out a suitable structure for doing so. This is a 'sheep and goats' essay title: it enables your tutor to sort the strong from the weak, the wheat from the chaff. Lots of gormless goats will opt for it because they think they can't go wrong, and will get poor marks. Don't do it (if you have a choice) unless you are sure you are a sharp sheep. If you don't have a choice and have to do this essay, be sure to recognize the trap that has been laid for the unwary.

# What is an essay plan and how will it help?

A pile of bricks and the knowledge that your final structure is going to have two stories and a roof is not enough. You need a detailed plan.

**If you are building a house, you need architectural plans. If you are writing an essay, you need an essay plan.**

There are two main reasons for having a plan.

First, the plan will help you to put your material down on paper in the right order without missing anything out and without putting anything in twice. Like the builder with bricks and roof tiles, you will have a blueprint that tells you where to put each idea or sentence as you work your way through the project. Don't lay your roof tiles in the basement, and don't put all your window frames into the ground floor walls because although they may look fine there you won't have any left when you start work on the upstairs.

Second, the act of producing a plan is a major stage in the design of a good essay. In putting together your plan you can try out different ideas and see which works best. You can see where there appear to be gaps (such as assertions without evidence, or evidence without sources) and make sure you fill them before you write the essay. Before you start writing the essay itself you can check that what you plan to write actually does what the assignment requires. If your plan doesn't work, your essay won't. It is better to spot problems on rough paper at the planning stage than to discover them embedded in the framework of the finished essay.

# How do I produce an essay plan?

There are lots of different ways to produce essay plans, and each person has their own preferred method. Usually the best way is to start with a simple statement of the question and the answer, and then expand outwards from there to include the reasoning, evidence and illustration that provide the structure of the essay.

## Step 1: Write down the question and the answer

You went through the process of producing a one-line answer to your question in Chapter 6. You then went away and made sure that you knew enough about the topic to be confident that your answer made sense. To start your plan, write down your question and your one-line answer once again. Box 7.5 offers some examples.

## Box 7.5 Examples of one-line responses to geography essay titles

Q: Why is glacier surging a geomorphological issue as well as a glaciological one?

A: Because surges affect landscapes as well as glaciers.

Q: Is 'health for all' an achievable goal for urban areas in the developing world?

A: Yes, but health provision would have to challenge the factors that focus resources on richer urban dwellers.

Q: What determines the speed at which glaciers move?

A: Key controls are mass balance, gradient, thermal regime and ice rheology.

Q: Evaluate how appropriate it is for geographers to use the nation state as the basic unit of territorial organization.

A: It is convenient for accumulating data, but differences in the characteristics of nation states make many types of comparison difficult.

Q: Discuss the importance of meltwater in glacier systems.

A: The importance of meltwater varies between glaciers in different environments, and meltwater can play a major role in both glacier behaviour and glacial geomorphology.

If you are not yet confident about your answer (or if you just don't know what the answer is) there are two approaches. The recommended approach is to go back one step and work on your material a little longer before settling down to write the essay. Go back to Chapter 6. An alternative approach is to plough on regardless and try to use the planning process as a way of organizing your thoughts and material to arrive at an answer. This is not recommended, as it can be a very time-consuming way of producing a plan, but if you are stuck for an answer it might be worth a try. If you use the plan to reach the answer, rather than vice versa, then you should take the answer you get and work forwards from that answer to a new plan. Your essay plan should be based on the point that you are trying to communicate, not the other way around. You don't draw up the plans for a new building before you know the purpose of the building you are going to produce! Doing the plan before you know the answer is like designing a building before you know what it's for.

**When you have your answer, write it down at the top of your plan.**

## Step 2: Draw your answer as a diagram

For most people, it is useful to convert their plan into a diagram for the early stages of the planning process. This is much easier than it sounds, and it helps you to design a good clear structure for your essay. Why do you think architectural plans are annotated diagrams, not just words? Why should your essay plan be any different? Try it. If you really don't like the diagrammatic approach, work through these steps anyway but apply them to your text-based plan. The steps are still valid.

The first jobs are to convert some of your text into symbols (such as ticks, crosses and question marks), to put key items in boxes, and to link the boxes together (or not, as appropriate) with lines, brackets and arrows. You can do this for the original question, as well as for your answer. For example, the question 'What causes pingos?' can easily be drawn as '?_Pingos' and immediately it is apparent that the answer will take the form: 'A_Pingos, B_Pingos, etc.'. We can now draw our answer with 'Pingos' in a box in the middle of the page (where else, for the central topic of the question?), and the various things, or combinations of things that cause them, off to one side of the page, connected to 'Pingos' by directional (causal) arrows. At this stage, make sure that your diagram matches your one-line answer. If it doesn't, adjust one or the other (or both) until they do match. Already you have the basis of a simple plan. Keep going.

## Step 3: Elaborate your plan with detail

Our plan so far is extremely basic, but it helps to show us where we need more detail. In our pingo essay for example, we have listed four possible causes of pingos. Immediately we are driven to wonder whether four is the right number. Should there be three, or five? Next we might wonder whether the items in our list have any order of priority. Are some more important or more common than others? Perhaps we should re-order the list. Now we see a paragraph order for our essay emerging. In any geography essay we have to wonder whether a simple cause and effect (A_Pingos) is universally true or whether it depends on some additional circumstance. In one environment 'A' and 'B' might be the dominant causes of pingos, while in another environment the dominant causes might be 'C' and 'D' . We need to subdivide our list into sections for different environments: causes A and B in one group (draw a loop around them) and causes C and D in another. So far we have arrows going from each cause to the feature, but in most geographical situations relationships are more complex than this. Perhaps some of the causes in our list interrelate (link them with arrows indicating the direction of any causal relationship). Perhaps there is a two-way relationship between pingos and one of the items in our list of causes: e.g. 'A' may cause pingos but also be more likely to exist where pingos already do. The arrow from 'A' to 'Pingos' needs to be double-headed.

At this point the plan will be looking a bit messy if you've been making all these changes on scraps of paper. Draw up a neat version so you can see what you have.

## Step 4: Organize the plan into sections

In many plans an order of play will already have become clear at this stage. For example, if you have a prioritized list at the core of our structure, then the order of play will be given by the order of priority. If no order is self-evident, you need to think about it now. Which bits of the plan would you explain to somebody first if you were talking them through your essay, and which bits would you save till last? Normally, items that sit close together on your diagram will go close together in your essay, so you can put colour-coded rings around areas of your plan and label them 1st, 2nd, 3rd, etc. If you find that bits of each group are scattered all over the picture, you might want to think about whether you have got things in the right places.

This is the stage when you should really be able to see if your plan is making sense. The structure of the essay is pretty much laid out now. If things don't seem to be in the right places now, you need to fix them because it won't get any better unless you do. From here onwards we're adding flesh to these bones, so you want to check that the bones are of the right sort of creature.

Eric says: *My plan is in trouble: I have to fit the skeleton of a rhinoceros into the skin of a gnat.*

## Step 5: Add evidence and illustration

If the bones of your structure come from the key points of your answer and the argument that links them, then the muscles and ligaments that hold the skeleton together are provided by the evidence that you present to justify your points. The essay will collapse in a nasty pile of goo if the evidence is not put in at the right places, so you need to test the placing of your evidence on the plan. The main types of evidence in most essays are references to published research and examples or case studies of the features you discuss. As a general rule of thumb each point that you make in your essay needs to be backed up by one or both of these types of evidence.

**On to your plan you should now add at least one reference and one example for each key point of your essay.**

At this stage you can also identify the intended locations of all the other features that you will incorporate into the essay, such as diagrams, tables, maps or useful quotations. It is sometimes useful to write these on to the plan in a different colour so that they show up clearly amid the detail. It's useful to see how your illustrative material is going to be distributed.

## Step 6: Fill in gaps and prune out dross

Your plan now should give a clear picture of exactly what the structure of your essay will include and where everything will go. If you threw on an introduction and a conclusion you would have a draft of the whole essay! This means that you can now see, in a clear diagram, where there are gaps in your evidence, where you have superfluous repetition of diagrams, and where you have too much space given over to a minor point or too little to a major issue. Take time now to work on the structure revealed by the plan. Fill in the gaps and prune out the dross. We are coming to the end of the planning stages, and getting close to starting the final essay.

## How do I go from the plan to the essay?

Some people like to go through an intermediate stage of converting their plan completely back into text form before using it as the basis of their final essay. Others prefer to work directly from the diagrammatic version. One advantage of re-writing the plan as text in the form of a hierarchical list of notes is that it forces you to think very carefully once again about the order that you decided on earlier in the planning. Your plan may have become a very complex diagram with bits and pieces linked together, but your essay has to be written (and read!) as a linear narrative. You have to start at the beginning and end at the end. You can't ask your examiner to jump to page four and read a bit and then come back to page two. Well, you can, and in some annoying exam scripts we've read that's exactly what we've been asked to do, but it generally loses marks.

Whether working from text or from an annotated diagram, the usual approach to using the plan is simply to transfer each section of the plan on to paper in the proper order, adding in all the detail that you identified during the planning process. In an exam you will not have time to do a complete rough draft, and you will have to go straight from the plan to the final version in one go. This makes the plan even more important in an exam than in a coursework essay. Look at Chapter 16 for more advice about dealing with an exam. For coursework assignments you should take the time to go through several drafts, the first of which will be this text conversion of the plan. At a first draft this can be done quite roughly, without worrying too much about niceties of layout, style or presentation. One danger is that you might end up with a string of material that looks like an undivided line of sausage. This means you have lost the section boundaries that you had in the plan. You need to rediscover those sections.

Eric says: *My first draft is an undivided stream of consciousness. What do I do?*

## How do I divide the essay into sections?

In your diagrammatic plan, the different sections of the essay were quite clear because they occupied different parts of the diagram or had boxes drawn around them or were drawn in different colours. In your first text version these divisions may be harder to see unless you were careful to draw a line (or start a new sheet of paper) between each section. The separation of the text into structural sections is the fundamental mechanism by which you give physical structure to your essay.

**You need to make sure that the sections you identified in your plan are clearly visible in your text.**

For now you can do this by drawing lines, or just leaving big gaps, between sections of text. If you were still working on our pingo example, you would have four main sections: an introduction (not yet written), the section covering causes A and B; the section covering C and D; and the conclusion (not yet written). Each of these sections will eventually comprise a number of paragraphs, each made up of several sentences. Most plans do not go down to the level of detail of individual sentences. It would be a good thing if they did, but we are not going to be completely unrealistic about this so we will save our discussion about how you organize paragraphs within sections of the essay until we get to the part where you are actually writing your essay. If you are one of those students who wants to plan right down to sentence level, feel free to read ahead through Chapter 8 before completing your plan.

## 'Finish' the essay before you start writing!

Before you get out that clean sheet of paper or open up that new word-processor file to start the serious version of your essay, you should have in plan form a 'finished' version of your essay. What this means is that you won't be searching for a theme, or surprising yourself with new information, or finding out what the answer is halfway through the essay. You have got all those things over with at the planning stage. All that remains now is to transfer your answer with all its attendant detail into the final essay. There is a lot of work to be done in the writing, but at this stage all the finding out and deciding should be finished. If it isn't, then finish it now before you proceed.

### *Conclusion to Chapter 7*

- A clear, logical structure is essential to a good essay. Both the logical components and the physical components of your essay need to fit together properly and arrive in the right order.
- Different types of essay require different types of structure, and you need to make sure that you are using an appropriate structure for your essay.
- Careful planning is the best way to develop a sound structure, and you must produce a detailed plan before you start writing the essay.

# 8

# How do I produce a well-written essay?

## *Chapter summary*

**If your essay is not well written it will not be clearly understood and you will not get a good mark. To produce a well-written essay you have to master technique, style and organization. You have to worry about the rules of grammar, the conventions of writing within your particular discipline, and the need to arrange material sensibly into fluent paragraphs and sentences.**

## What do tutors mean when they talk about 'well-written' essays?

The main criteria by which you can judge whether an essay is well written are technique, style and organization (Box 8.1).

### Box 8.1 Criteria by which to judge whether an essay is well written

*Technique*

The writing must be technically correct in terms of grammar and syntax. This covers things like punctuation, word order, agreement of tenses, and use of verbs. Errors make the essay harder to read and sometimes difficult to understand. Errors make you look incompetent.

*Style*

The essay must be written in a style appropriate to the nature of the assignment. Typically you should be aiming for the formal, scientific, technical and professional, not the casual, informal or sensational. Clarity and precision are key targets.

*Organization*

The essay needs to be organized into paragraphs, and the paragraphs into sentences. The organization of these units should reflect the structure of the essay such that structural breaks correspond with organizational boundaries. Organization is the visible manifestation of structure.

When we talked about essay structure in Chapter 7 we used the analogy of an architect designing a building. If the building was well structured, things would be in the right places and the building shouldn't fall down. Continuing that analogy we could say that if the architect is responsible for designing a good structure, then the builders and decorators are responsible for ensuring that the final construction is well built. They actually cement the bricks together, plaster the walls and paint the woodwork. If the quality of workmanship at this stage is poor, then the final product will be unsatisfactory, despite the good work at the planning stage. If the builder doesn't keep the rows of bricks perfectly level, the plasterers make the walls lumpy, and the decorators let drips set in the paintwork, the building will become the architectural equivalent of a poorly written essay.

**Poorly written essays arise from lazy, careless or unskilled work.**

With your essay, even if you have a good plan and a sound structure, you can't afford to be lazy, careless or unskilled when it comes to writing the essay itself. Even if the design was sound, poor execution will let you down.

Eric says: *My tutors always say: 'Your essay was not well written.' What do they mean, and what can I do about it?*

## Technique: grammar, syntax and all those rules

Trying to communicate with someone who doesn't speak your language is just like trying to play a game with somebody who plays by different rules. You won't get very far. Languages have rules so that everybody who uses the language uses it in the same way. If they don't, communication breaks down. Grammar and syntax are the rules that everybody has to follow to make sure that we all use English in the same way and to make sure that we all understand each other clearly. Incorrect use of English is OK when you are dealing with somebody face to face in day-to-day business, because meaning is conveyed partly by non-verbal means and by the context of the situation. If I fill my car with petrol and then walk up to the cashier with money in my hand, it doesn't need correct syntax to be sure that the cashier will understand what I want to do. In an essay you don't have the benefit of face-to-face contact, you are not dealing with an 'obvious' transaction, and you are trying to communicate complex sets of ideas in a precise way to a reader who has nothing to go on but the words you have written down. You need to be absolutely sure that you've written them down in such a way that they mean exactly what you intend, in exactly the same language that the reader would use. In other words, the writer needs to be working from the same rule book as the reader. Fortunately the rule book is widely publicized and well known. It's called 'English'. Use it wisely.

This book is not the best place to go if you need lessons in grammar. Box 8.2 gives you some examples of the most common abuses of English that tutors weep over. If you think you may be guilty of these, seek help. Your institution almost certainly has people who can help you to improve your English language skills, whether or not English is your first language. Don't be afraid to ask. Your tutor should be able to point you in the right direction.

### Box 8.2 Some examples of common abuses of English over which tutors weep

*Wrong:* The number of students are increasing.

*Right:* The number of students is increasing.

*Wrong:* There were less people at the lecture than I'd expected.

*Right:* There were fewer people at the lecture than I'd expected.

*Wrong:* Eric never gets things done; he always prevaricates.

*Right:* Eric never gets things done; he always procrastinates.

*Wrong:* The growth of plants is effected by the climate.

*Right:* The growth of plants is affected by the climate.

*Wrong:* The affect of climate on plants is to modify their growth rates.

*Right:* The effect of climate on plants is to modify their growth rates.

*Wrong:* Its never too soon to start work on your essay.

*Right:* It's never too soon to start work on your essay.

*Wrong:* An essay can be judged by the quality of it's introduction.

*Right:* An essay can be judged by the quality of its introduction.

*Wrong:* The data is flawed.

*Right:* The data are flawed. ['data' = plural]

## Style: the appropriate, the inappropriate and the ugly

Different types of writing need to adopt different styles. The formal and technical style that would be appropriate to a scientific paper would not be suitable for an article in a tabloid newspaper. The parish newsletter needs to be written in a different style from a police report. Your essay needs to be written in a different style from a letter to your mum. The reasons for this are similar to the reasons we gave for following the rules of grammar. You need to speak the same language as your reader and you need to make sure that your writing adequately communicates your meaning. This involves following certain conventions that apply to your particular type of writing. In writing an essay you are trying to demonstrate your academic competence and communicate complex ideas with maximum clarity and precision. To achieve that, you need a clear and precise style and you need to follow the conventions of the academic world. You don't need to sensationalize like a newspaper to make issues seem more exciting than they are. You can't get away with ambiguous or sloppy phrasing like you would in a note to a friend. You can't dedicate lengthy sections of text to entertaining but non-essential issues as you might in writing a novel.

Key style points for academic writing include the following.

- Be clear.
- Be precise.
- Be accurate.
- Be detailed.
- Be unambiguous.
- Be succinct.
- Be formal.

Some of these take a lot of thought and practice, but others are easy to achieve just by following some simple do-and-don't guidelines. For example, you can go some way to ensuring a formal style in your writing by following the guidelines in Box 8.3. Your essay will immediately lose its air of formality if you drop in words of slang (the plan is nuts), inappropriate use of the first and second person (I, we, you), or casual abbreviation (there's, instead of there is). Be careful not to write in the same style that you speak. Everyday speech is too informal for most essay-writing situations.

There can be exceptions to these rules, of course. For example, it might be acceptable to use 'I' in your answer if the question specifies 'you'. Nevertheless, even where the question gives you an excuse to write 'I think this or that', it is nearly always better to write in impersonal terms. If you make an assertion without crediting it to a specified source then it will be quite clear to the reader that you are giving your own opinion. After all, the reader knows quite well that it is you who wrote the essay, so you don't need to write 'In this essay I will show . . .'. It is better to write 'This essay will show . . .'. Your conclusion should not say 'I think the most important factors are . . .'. It is better to write 'The most important factors are . . .' and trust the reader to understand that this is what you think.

### Box 8.3 Simple rules for ensuring an appropriately formal style for essays

**Don't write in the first person**

*Instead of:* 'I think there are three main causes . . .',

*use, for example:* 'Three main causes may be considered . . .'.

*Instead of:* 'If it rains you know the ground will get wet',

*use, for example:* 'Rain causes the ground to get wet'.

*Instead of:* 'We will be in trouble if there is an earthquake',

*use, for example:* 'An earthquake would cause problems.'

**Don't use casual abbreviation**

*Instead of:* 'there's', 'wouldn't', 'can't', 'don't',

*use:* 'there is', 'would not', 'cannot', 'do not'.

**Don't use clichés or slang**

*Instead of:* 'Glaciers usually move dead slow',

*use, for example:* 'Glaciers usually move very slowly'.

*Instead of:* 'The President got a pretty cool reception',

*use:* 'The President was met by a hostile crowd',

*or:* 'A cold buffet was organized to celebrate the President's arrival'.

## Organization

An essay should not be presented as a single lump of text running unbroken right through from the top of the first page to the end of the essay. It needs to be divided into sections. When you created your essay plan you devised a structure that identified several different parts to the essay. At its simplest, this structure involved an introduction, a conclusion, and a list of several items in the middle. These separate sections in your structure need to appear as different sections in the essay.

**The basic instrument that you should use to organize your text into sections is the paragraph.**

Each section should consist of one or more paragraphs, and each paragraph in turn comprises a number of sentences. If you enjoyed our architectural analogies, you might like to think of paragraphs as the rooms in your building. The introduction is the entrance hall. Sentences within each paragraph are the furnishings and colour scheme that give the room its form. Paragraphs and sentences are so important to the production of a well-written essay that they deserve sections of their own in our chapter.

## Paragraphs

A paragraph is a free-standing, self-supporting unit of text within your essay. In a nice straightforward organization, a paragraph will equate to one major point or item from your essay structure. For example, if your essay structure is a simple list of five items your essay could consist of a paragraph of introduction, a paragraph for each of your five list items, and a paragraph of conclusion. That's a pretty basic approach, but it's a lot better than some of the essays we've marked over the years.

Students often go wrong by misjudging how much material should go into each paragraph. Tutors hate to see essays where page after page is peppered with tiny little paragraphs just a line or two long. They hate just as much to see an essay where single paragraphs stretch over several pages. A paragraph needs to be long enough to do its job, but no shorter and no longer. What is its job? The purpose of a paragraph is to communicate a complete point, and a complete point typically includes at least the following:

- **a core statement** of information or opinion;
- **a context** to establish the significance of this statement to the point of the essay;
- **evidence** to support the statement;
- **examples** to illustrate the statement and/or the evidence;
- **references** giving the sources of the evidence, examples and other material used.

Even this very basic set of requirements shows us that an average paragraph is going to have to contain several sentences. We are probably looking at something of the order of about one hundred words minimum for the basic common-or-garden essay paragraph. Box 8.4 gives an example of a paragraph of this type.

This very simple type of paragraph is very useful in straightforward situations and can serve as a starting point from which to build more complex paragraphs if the need arises. It does its job of making and supporting a point within the essay, and it doesn't do anything really to annoy the reader, so you need to make sure you can write paragraphs like this. Simple though it is, it is better than many students manage when they start writing essays. Here are some simple rules for paragraph construction that avoid the most common errors made by students.

- Always include the full complement of point, evidence, example, context and references.
- Avoid one-sentence paragraphs.
- When you've made your point, stop and move on to the next paragraph.
- Make sure your paragraph has a beginning, middle and end.
- Make sure your context signposts the reader's route through your material.

## Box 8.4 An example of a very simple common-or-garden essay paragraph

*This is part of an answer to the question 'Examine the concept of a glacial sediment budget'.*

'The debris that many valley glaciers deposit in ice-marginal moraines can be derived from both subglacial and supraglacial sources. These moraines therefore provide important evidence about sediment routing through individual glaciers and facilitate quantification of the sediment budget. Debris in moraines typically includes both angular, frost-shattered fragments characteristic of supraglacial transport and sub-rounded, faceted, striated clasts that show the effects of subglacial transport processes (Smith, 1985). Brown's (1997) study of moraines deposited by 12 Swedish valley glaciers indicated that the ratio of supraglacial to subglacial debris ranged from 1.5:1 to 6.3:1 and depended largely on the relief of the valley sides.'

*Sentence 1* is the core statement.

*Sentence 2* gives context.

*Sentence 3* provides evidence (with a reference).

*Sentence 4* supplies an example (with a reference).

Of course, a set of rules cannot apply universally to all situations. What we have tried to do is show you a safe basic approach to constructing sound paragraphs. When you know how to do that, you can start to experiment with different approaches. Intelligent individuality is the hallmark of a great essay. Don't allow our petty rules to stifle your imagination. However, if you experiment with one-line paragraphs or try making points without the use of evidence and your tutor lands on you like a piano dropping from a fourth-storey window, don't say we didn't warn you.

Eric says: *Intelligent individuality is the hallmark of a great essay. Discuss (intelligently).*

### Additional notes for advanced users of paragraphs

We've said that paragraphs are free-standing units of text. That doesn't mean they are independent of each other. Going back to our architectural analogy, if you went into the first room in a house and it was furnished in Louis XVI antiques and then walked into the next and found it was 21st century minimalist, you'd be shocked. You expect some continuity of style in a house. And it's the same with an essay. You need to make the transitions from one paragraph to the next smooth and harmonious. Sometimes this is easy (see Box 8.5), but at other times you have to work a bit harder.

### Box 8.5 Some paragraph transitions

*Essay title*

What are the main factors controlling soil formation?

*First paragraph*

Several soil scientists have discussed the primary controls on soil formation . . . [paragraph summarizing primary controls] . . . Jenny (1941) emphasized the importance of climate, parent material, relief, vegetation and time.

*Next paragraph*

Climate is important to soil because . . . [paragraph about climate] . . . Climate is therefore very important. However, its effect depends to a large extent on the parent material.

*Next paragraph*

The role of parent material is . . . [paragraph about parent material] . . .

*Etc.*

## Sentences

Sentences are what paragraphs are made of. If the purpose of a paragraph is to make a complete point with all its attendant detail, then the purpose of each sentence is to deal with one part of that assembly. Sentences can be simple. That sentence was very simple but this one, which includes several sub-clauses, is more complex and therefore, notwithstanding the intelligence of most of our readers, a little harder to follow. In general, you should tailor your sentences to be as easy as possible to understand. You can avoid long, complex sentences by making each different part of the sentence into a complete sentence of its own. Consider this revision of our earlier 'complex' sentence into several shorter sentences. 'That sentence was very simple. Our earlier example was more complex. It included several sub-clauses. Notwithstanding the intelligence of our readers, the complex sentence was harder to follow.' Much easier.

'Much easier' is not a sentence. Sentences have to follow the rules of grammar and syntax that we referred to earlier. Box 8.6 lists and illustrates some of the basic rules that sentences have to follow.

**Box 8.6** Some of the basic rules that sentences have to follow, and examples of the non-sentences that result if you break the rules

**Sentences must have a verb.**
*Wrong:* 'Also France.' 'For example, Snowdon.' 'In the interests of public safety.'

**Tenses and cases must agree.**
*Wrong:* 'They will came last Tuesday and eats a meal.'

**The subject must agree with the verb in person and number.**
*Wrong:* 'I comes, he come, and they comes.'

If these rules don't make perfect sense to you, or if your tutor writes comments like 'This is not a sentence' on your essay, you should seek help.

### Additional notes for advanced users of the sentence

Getting sentences 'right' is one thing, but making them really good is another. If you are struggling to make your English comprehensible, then getting your sentences right should be your first target, but if you can already do sentences you might want to develop your technique further to make your sentences really work for you. Adequate sentences get your message across. Really good sentences make your reader enjoy getting your message. So far our chief goal has been clarity. Let's now add variety and fluency.

The short sentence packs punch. Longer sentences, involving sub-clauses, tend to slow things down. You can use variation in sentence length to control the rhythm of your paragraph and thereby reinforce the message that you are trying to convey. Imagine trying to communicate a list. You need to be clear. You need to be direct. You need to be accurate. Those three short sentences highlighted the fact that we were listing three individual items, while the greater length of this sentence draws attention to the fact that we have finished the list and entered the discussion phase. A short sentence next would sound good. Even without content, the composition of text can make a difference to the readability of your work. Think about it. (See how we finished the paragraph with a short sentence intended to pack punch?)

Eric says: *Paragraphs in which all the sentences are the same length tend to sound dull. Paragraphs also sound dull when the same word is repeated in adjacent sentences. If the word 'dull' appears again in this paragraph it will be a dull paragraph. It was.*

# How do I include quotations?

There are several special circumstances that you need to accommodate in your writing. One of these is the insertion of references (Chapter 12) and another is the use of quotations. A quotation is where you write down exactly what somebody else has said or written previously. If you report it in your own words rather than copying it down word for word from the original, it doesn't count as a quotation and you can incorporate it into your text as part of a normal sentence. For example, you could write: 'Smith (1992) argued that further development would be unsustainable.'

However, if you wanted to report exactly what Smith had said, you would need to quote verbatim from the original source. To do that you have to put the original words within quotation marks and provide a reference to the specific page on which these words appeared in the source. For example, you could write: 'Smith (1992, p. 27) wrote that "further development in this region will not be sustainable beyond the end of the decade".'

If you are presenting an extended quotation that runs to one or more complete sentences it is conventional to put the quotation as a separate, indented paragraph. You might also put the text of the quotation in italics to make it stand out clearly from the rest of your text. For example, we might quote what we wrote about presentation in our book about geography dissertations:

> *'Errors of spelling, punctuation, syntax or grammar will make your report more difficult for the examiner to read. This will irritate the examiner, and may make the examiner angry. An irritated or angry examiner is probably going to be a mean examiner.'*
>
> *(Parsons and Knight, 1995; p. 133.)*

Quotations and other special circumstances might be the subject of specific instructions in your institution's guidelines for producing essays. Check, and make sure you do what is required.

# Signposting

One of the characteristics of a well-written essay is that the reader never gets lost. In a badly written essay, the reader sometimes loses track of the argument, loses sight of what point a particular example is supposed to be illustrating, or has to go back and re-read the start of a paragraph in order to understand the significance of the end of the paragraph. These are all symptoms of poor signposting. It is the job of the writer to make sure that the readers know where they are, where they have come from and where they are going as they read through the essay.

Eric says: *If you get lost reading your own essay you really need better signposting!*

Consider this short section of text:

> *In very dry environments, even where temperatures are low, little snow can accumulate. In areas with long warm summers, even large winter accumulations of snow melt during the summer. Aspect is important because it exerts a major control on local variations in temperature, precipitation and snow accumulation.*

That was OK, but what on earth is the point of the paragraph? What is the writer getting at? Where are we going with this story? It is not clear. However, try re-reading that text with one extra sentence at the front of it:

> *The distribution of glaciers is controlled largely by temperature and precipitation. In very dry environments, even where temperatures are low, little snow can accumulate. In areas with long warm summers, even large winter accumulations of snow melt during the summer. Aspect is important because it exerts a major control on local variations in temperature, precipitation and snow accumulation.*

That one additional sentence enables us to make sense of all the rest. It serves as a signpost telling us where we are in the landscape of the essay. (If you are still using our architectural analogy, and you have built yourself a multi-storey department store, then this is the voice in the lift that says 'Second floor, ladies' and gents' underwear, going down.')

This is more or less the same point that we made when we talked about the need for context in each paragraph. Context is one example of signposting in an essay, as it provides obvious indicators to the reader as to what is coming up next. These signposts need to be placed regularly throughout the essay, possibly even at the start and end of each paragraph if things are looking complicated.

The most important bit of signposting is right at the start of the essay, in the introduction, where you have to tell the reader what is coming up in the whole essay. That's what we'll look at in the next chapter, as we move on from the planning stage and start writing our essay.

## Conclusion to Chapter 8

- A well-written essay is an essay that obeys the rules and conventions of technique, style and organization.
- A well-written essay is not marred by errors of spelling, punctuation or grammar.
- A well-written essay is easy to read and will allow the marker to see clearly all its strengths and virtues.
- Lazy, careless and unskilled students will produce badly written essays and they will be penalized accordingly.

9

# What goes at the beginning of the essay? The introduction

## Chapter summary

A good introduction will show how you have interpreted the question and how you are going to present the answer. It will help the reader to get the best out of your essay, and it will give the reader a good first impression of your knowledge, understanding and essay-writing skills. Do not waste time warming up, but get straight down to business and start communicating important ideas and information. There are a lot of advantages to putting your answer right at the front of your essay, but you also need to indicate that the evidence for the answer, and discussion of all the points for and against, will be coming up in a logical sequence in subsequent paragraphs. Your introduction should develop directly from your plan, and should serve as a signpost for you as you write the rest of the essay.

## How important is the introduction?

Very. Your essay must have one, and it must be a good one.

Eric says: *Your tutor reads the introduction first. First impressions are important.*

## What is the introduction's job?

The introduction does several important jobs. At least, it does if you write it properly. If you don't, it doesn't, and your essay is in trouble right from the start. The main functions of the introduction are as follows.

- It indicates how you have interpreted the question.
- It sets out the main themes of your answer.
- It demonstrates the overall structure of your essay.
- It lets the reader see what issues are going to be covered in the essay.
- It shows that your essay is relevant to the title.
- It establishes the clear and professional tone of the essay.

- It shows that you can see through the detail to the core of your argument.
- It convinces the reader immediately that you know what you are talking about.
- It convinces the reader immediately that you know how to write an essay.
- It enables you to check what you are doing before you get too far.
- It serves as a point of reference to keep you on target as you write the essay.

If your introduction is no good, the following serious problems will arise.

- Readers won't be sure how you have interpreted the question.
- Readers won't have a clue what is coming next in the essay.
- Readers won't know what you think are the main issues surrounding the topic.
- Readers will think you are incompetent.
- Readers will consciously or otherwise have bad essay marks forming in their minds.

## What should I not waste time on in the introduction?

The tutor's heart sinks when, in the first few moments of reading an essay, it becomes immediately apparent that this is a poor introduction and therefore, probably, heralds a poor essay. There are several tell-tale signs that an experienced tutor learns to fear, and they are symptoms of common weaknesses that affect many essays. Learn not to upset your tutor. Make sure your introduction avoids these pitfalls.

The introduction is not an opportunity for you to 'loosen up'. It is not there for you to make sure that your pen is running smoothly, or that your keyboard is free of dust. Every second of the reader's time that passes in the introduction without you scoring effective points is a second when the marker will be registering 'missed points' on the mental score sheet. Get in there fast and start making crisp, worthwhile points immediately.

Unless the essay title involves unusual or ambiguous terms, you do not need to waste time defining words from the title at the start of the introduction. Please, please do not start your essay with the words: 'The Oxford English Dictionary defines . . .'. Your tutor's heart may break.

Your introduction will rightly tell the reader what you are going to cover in the essay. It should not tell the reader what you are not going to cover. If it's important it should be in the essay. If it's not, it has no place in the valuable space of the introduction.

## Should I answer the question right at the start of the essay?

1. This is a question that usually crops up quite early when people get down to arguing about how to produce a good introduction, so let's get it out of the way before we start producing our model examples. You will notice that in this paragraph we have not yet told you what we think the answer to the question is. Irritating, isn't it? If this were an essay we could now say that there are lots of issues to consider, that the essay will deal with them each in turn, and that we will arrive at an answer when we get to the conclusion in about 2000 words time. Does that help? No? Try the next paragraph and see if you like it better.

2. It is nearly always best to provide a basic answer to the question in your opening paragraph. It demonstrates to the reader that you have understood the question, that you have something worthwhile to say in response, and that you are focused on the relevant issues. By including some elaboration, as we just did in that last sentence, you can also signpost the direction that the rest of your essay will take. If this were our introduction you would be able to predict by now that the core of our essay would comprise detailed analysis of those three main reasons that we listed in the second sentence of this paragraph. We should also mention that a minority of tutors may disagree with our answer, so the reader will know that there is a discussion of that controversy coming up later. Compare this second paragraph with the one above in relation to the list of bullet points at the top of this chapter, and decide for yourself whether it was useful to include a basic statement of our answer at the front of our 'essay'.

Eric says: *Should I make my point and then back it up, or should I back up my point and then make it?*

We have known tutors who prefer the answer to be saved to the end of the essay, and who criticize students who reveal it too soon. They use the 'cart before the horse' analogy, and say that you should not reveal the answer until you have presented all the evidence. We think they are wrong. It is true that you can't demonstrate your answer without evidence, but you can tell the reader what your answer will be before you show them the evidence and discuss the pros and cons of your point of view. You can announce that you are about to deliver a cart, before you show us the horse that will pull it. Look at the different ways of answering questions in Box 9.1.

## Box 9.1 Different ways of answering questions

**Party question:**

'Can you come to the party tonight?'

*Straightforward answers:*

'No, I can't come to the party because my parents are visiting and I have to go out with them tonight instead.'

'Yes, I can come because although I am going out with my parents I can make an excuse and leave them early.'

*Less straightforward answer:*

'My parents are visiting. I have to go out with them tonight. This means I am not free to come to the party, so no, I shouldn't come. However, I could make an excuse and come anyway.'

**Geographical question:**

'What is the capital of Greenland?'

*Straightforward answer:*

'The capital is Nuuk. It used to be called Godthaab, but the name was changed after Greenland established Home Rule and cast off a lot of its Danish colonial inheritance.'

*Less straightforward answer:*

'The capital used to be called Godthaab. That was a Danish name. Greenland now has Home Rule, and place names are reverting to Greenlandic versions, so the capital now has a different name, but it's the same place. It's called Nuuk now.'

An essay that puts a summary of the answer in the introduction is like a TV detective show that lets you see the crime being committed at the beginning so you know who is the guilty party. As the programme continues you are then in a position to judge how good a job the detective is doing as he follows good leads and bad in the investigation. You are in a position to judge the evidence, just as the reader of an essay wants to be in a position to judge the arguments presented in it. By contrast, if the essay does not announce its key point at the start, then the reader is in the position of a viewer of a mystery drama in which the solution to the puzzle remains hidden till the final scene. In that show, the viewer is led along blind alleys, follows red herrings, and is as puzzled and confused as the fictional detective trying to solve the crime. This may be entertaining, and it may allow the horse (the evidence) to go before the cart (the answer), but it doesn't enable the viewer to understand and judge each piece of evidence as soon as it is presented. When you surprise the reader with the final twist in the conclusion, do you really think your examiner will have the time or inclination to go back through your essay to re-examine the bits of evidence that finally turn out to have been significant? Probably not.

**It would have been better to have signposted the bits of evidence very clearly when they first appeared.**

In our 'less straightforward answer' for the party example in Box 9.1, the significance of the fact that parents were visiting was not clear until the end of the third sentence. How many of you turned off after the second sentence of the less straightforward answer and missed the important point at the end? In an essay, where these would be pages rather than sentences, this would be a very bad thing.

If you are not sure whether you like the idea of putting the answer right up front or the idea of saving it up for the end, think about what the essay as a whole is trying to achieve. The essay is trying to communicate your answer to a question, or your opinion about an issue. You have a message of some kind to get across, and your essay is your means of communicating it. What matters is that you communicate it clearly and that the reader understands each point that you make. Look at the examples in Box 9.2 and decide which mini-essay does the job best.

Box 9.2 Two possible responses to the title 'What is the most interesting feature of the Ordnance Survey map of Warwickshire?'

*Which essay can you evaluate most quickly and easily?*

1. The map of Warwickshire is . . .*blah* . . . *blah* . . . *blah* . . . Thus the most interesting point is that all the towns on the map start with the letter W.
2. The most interesting point about the map of Warwickshire is that all the towns on the map start with the letter W. Towns in the . . . *blah* . . . *blah* . . . *blah* . . .

## What should I put in the introduction? The basics

An examiner ploughing through a pile of essays perks up and takes notice when an essay begins with a crisp intelligent introduction that addresses the question directly and signposts the essay clearly. What you need to put in the introduction, therefore, are sentences that do those jobs.

Your introduction should:

- present the brief answer to the question;
- present the list of key issues that signposts the structure of the essay;
- identify the broader significance of your answer;
- recognize the controversy or the alternative viewpoint that you will consider.

If somebody in conversation asks you a difficult question that demands a complicated answer, your immediate response would have to warn the listener that the answer has several parts, indicate the general direction the answer will take and convince the listener not to walk away before you finish speaking. If the question was 'What is the best thing about studying geography?', you might begin with something like: 'There are lots of good things about studying geography.' That gives the general tone of your response (positive and affirmative) but it doesn't actually answer the question or generate much interest. You could start with: 'Different people like different things about geography.' That puts a little more interest into the response, but still doesn't convince the listener that you have anything worthwhile to say. You could try: 'Some people think that the distinctive subject matter of geography is its best feature, while others think that the generic skills that it teaches are more significant and yet others argue that its greatest value is in the vocational training for professional geographers.' This opening line indicates that the answer is not straightforward, explains that different people have different ideas, and lists some of those different ideas which, presumably, you are now about to go on and elaborate. This opening line has explained that it's a complicated answer but also indicated the main issues that you are going to include in the answer. In conversation, most people can achieve this successful introduction time after time. In essays it should not really be any more difficult than this.

## How should I start the introduction?

Sometimes, putting pen to paper for that very first line is hard to do. If you are struggling to know quite how to begin, one safe approach is to get straight into the thick of things and answer the question. You already have your one-line response, because you had to make it up when you were starting the planning process, so, if in doubt, start with that. This will achieve several of the things that you need to do in the introduction.

- It will prevent you from wasting words with warm-up waffle.
- It will present your key point.
- It will demonstrate how you have interpreted the question.
- It will show the reader that you are getting directly down to business.
- It opens up the way for a logical follow-up sentence.
- It breaks open that scary clean white page.

If you can't even think how to word your one-line response properly, the emergency-start procedure is just to turn around the wording of the title to make it into the form of an answer. Box 9.3 shows how you can turn questions around into answers, and the 'answer' lines for each title in the box could serve as the first line of your introduction.

**Box 9.3 Examples of how you can turn around the title to generate the first line of your answer**

Q: What are the main goals of modern urban renewal programs?

A: The main goals of modern urban renewal programs are . . .

Q: Why do some rocks weather more rapidly than others?

A: Some rocks weather more rapidly than others because . . .

Q: Evaluate the success of town planning in the UK.

A: The success of town planning in the UK has been . . .

Q: Discuss the role of religion in the geography of crime.

A: The role of religion in the geography of crime raises several important issues . . .

## Where do I go with the follow-up sentence?

In the previous bullet list we said that putting the one-line answer at the front of the introduction would open the way for a logical follow-up sentence. From the various examples we have used in this chapter you will see that after we give an initial answer, we generally back it up with detail or justification, and sometimes add a limitation or an alternative viewpoint. We also try to explain why our answer is significant in the broader context of the topic. A typical format would be: 'The answer is *abc*, because *x* and *y*, although sometimes *p* and *q*. This is important because it affects *z*.' Try this example: 'It *is* useful to put the answer at the front *because* it helps you to make your point clear. *Although* some people think differently, we suggest that a direct approach to the question is *important* to keep your essay focused from the start.' So, your next few sentences after the answer need to cover those points. *Why* is that the answer? Is there an *alternative* answer? What *difference* does your answer make? Look at Box 9.4 for a couple of examples of this type of follow-up. You will see that we are beginning to produce a very basic but broadly sensible type of introductory paragraph.

**Box 9.4 Follow-up sentences for different one-line answers**

*Answer line:*
Some rocks weather faster than others because different rocks have different chemical properties.

*Follow-up line:*
However, weathering rate depends not only on rock type but also on environment. Identical rocks weather at different rates in different thermal and hydrological settings, so the prediction of weathering rates depends on environmental as well as lithological controls.

*Answer line:*
The main goals of modern urban renewal programmes are to provide high quality environments for economic, social and environmental purposes.

*Follow-up line:*
Specific programmes focus on the specific problems of individual cities, but common goals and approaches can be recognized across a broad spectrum of urban locations.

*Answer line:*
The role of religion in the geography of crime raises several important issues.

*Follow-up line:*
Many studies have focused on the importance of religion as a motivating factor in specific types of crime, and hence a controlling factor in the geography of crime. However, less work has focused on how different religious groups respond to crime, and on the impact of this on the initial reporting of religiously motivated crime.

## How do I finish the introduction?

You should try to give the introduction a self-contained structure with a beginning, a middle and an end. For the very simple model the beginning would be the 'turned around' question (the answer), the middle would be the detail, elaboration and counter-argument, and the conclusion could be the statement of significance.

Before you move on from the introduction, you need to make sure that it really does signpost the direction that the rest of the essay will take. If it doesn't already do that, then that is how you can finish it off: with a signpost. In the best introductions, however, the structure and content of the introduction is a signpost in itself. For example, in our simple model the order of material in the introduction is a microcosm of the structure of the whole essay. If the readers have followed the order of points in the introduction (here's the answer, this is why, here's an alternative, and this is why our answer is important) then they are all ready for an essay that is divided into four sections that: (1) explain why the answer is what you say; (2) consider alternative viewpoints; (3) explore the broader significance of the issue; and (4) conclude with a reminder of what your answer was in light of the evidence that you have now presented in the body of the essay.

**The structure of the essay should be entirely predictable from the introduction.**

The introduction is built as one big signpost. You don't need a clumsy extra signpost that says 'This essay will start by explaining why I think this is the answer, and then . . .'. However, if your introduction is not a model of clarity, then a clumsy extra signpost might be called for at the end. If you have felt the need to erect a big flashing neon sign at the end of your introduction, you may want to go back and see if your introduction could be structured more effectively. A big sign is a worrying sign.

If all is well and you don't need a big ugly signpost, then you have two options at the close of the introduction. One is to stop as soon as you have finished. Unless you have a good reason to do otherwise, this is usually a good idea. Another option is to add something a little bit special. This something must not detract from the directness and clarity of the introduction, and it must not be something that would have fitted perfectly well in one of the earlier sections of the introduction. It should be something that will make the reader notice that you are offering a little bit extra, and it should make the reader look forward with even greater anticipation to what is coming next. This is a neat trick, and something that separates the excellent introduction from the merely competent. It can be achieved in many different ways but it takes a smart student to pull it off well, so be careful.

Some advanced techniques include:

- the cleverly appropriate quote from an advanced source;
- the erudite comparison of the title of your essay with a slightly altered title that would have had interestingly different implications;
- the intelligent metaphor.

These can go badly wrong in unskilled hands, so be careful. Be especially careful of The Joke.

Eric says: *Your tutor almost certainly has no sense of humour. When your tutor says, 'This essay is a joke,' no one is laughing.*

## How long should the introduction be?

The introduction should be long enough to do its job. No longer. No shorter. If your introduction runs to fewer than three sentences you might want to check whether you have really covered all the points that we identified in those bullet points at the top of this chapter. If it runs to more than about 250 words you might begin to wonder whether you are taking too long to make your point. You should try not to let your introduction run over a page boundary, since the page boundary is also a psychological boundary for the reader. Once you turn the page, you feel that you must be into the depths of the essay.

### Can the introduction take up more than one paragraph?

If you are following our advice that a paragraph is a unit of text that makes a single complete point, and if you succeeded in synthesizing your essay into a one-line response as we recommended in Chapter 6, then you should find that your introduction turns out to be a single paragraph. A big advantage of this will be that the physical layout of the text on the page will help the reader to see where the introduction ends and where the next section of the essay begins. If your introduction runs to more than one paragraph the reader may mistake the end of the first paragraph for the end of the introduction, and your signposting will break down. There may be situations where a multi-paragraph introduction could be appropriate, but these will be the exception rather than the rule.

## When should I write the introduction?

We said at the start of the chapter that part of the job of the introduction is to enable you to check that you know what you are doing before you get too far into the essay and to serve as a point of reference to keep you on track as you work through the essay. It can only do these jobs if you write it before you write the rest of the essay.

People sometimes say that you should leave the introduction till last when you write an essay. This is only a good idea if you don't know what you are going to put in your essay before you start. In other words, it's only a good idea if you don't know what your answer is, don't have a clear plan, and intend to make the essay up as you go along. This is a bad thing to do. In fact, it's so bad that we feel pretty confident in saying that you should never do it. If you find yourself in a position where you want to leave the introduction till after you've written the essay, then your essay is probably in serious trouble and you should go back to the stage we discussed in Chapter 6, where you started to figure out what you would put into your essay.

**Even in an exam you should not start writing if you don't know what you are going to say in the end.**

You wouldn't start out on a journey from A to B unless you knew which direction B was in. It's the same with an essay.

It may well happen that while you are in the process of putting the essay down on paper you have an excellent new idea, or recognize some flaw in your original argument. In this situation you may need to make some addition or correction to your introduction so that the reader doesn't get a surprise when the essay suddenly lurches in an unexpected direction, or when you suddenly produce a surprise witness in the final scene. Your whole essay should remain open to modification and improvement right up to the last minute, and the introduction is no different. Nevertheless, you need to write a first draft of your introduction before you start the first draft of the rest of your essay, and you need to write the final draft of your introduction before you start the final draft of your essay. You can always come back and rework it a bit, but you have to get it out there in front to show you the way.

Here's the order of play so far, right from the top:

- Make sure you know what question you are doing.
- Figure out the answer.
- Design the overall structure and prepare a plan.
- Add detail to your plan, and think about it a lot.
- Write the introduction, and check that it matches your plan.
- Go ahead and write the rest of the essay (we'll discuss this bit next).
- Keep checking that the essay is following the signposts in the introduction.
- At the end, go back and check that the introduction still matches the essay.

You have to go back and check the introduction at the end because during the course of writing the essay, especially if it is a coursework essay that has been written over a number of days, you may have had new ideas while writing. That's OK. One of the great things about doing essays on a word processor rather than writing them by hand is that it is easy to make changes right up to the last minute. In an exam, this would get messy, so you have to plan even more carefully in advance.

## What if I can do all that? Additional notes for advanced users

So far we have explained the basic elements of an introduction. However, it would be a mistake to think that there is a single formula that will apply to all situations and get you a first-class grade. The approach we have given you here should certainly keep you out of serious trouble, but if you want to do really well you should build on this basic framework to develop introductions that are appropriate to the individual needs of each different essay that you tackle, that reflect your individual approach to the answer, and that allow you to set up more sophisticated essay structures. Ways of doing this will become easier for you to discover as you have more practice writing essays and as you read the advice about later sections of the essay in the next few chapters.

Eric says: *Don't just do what they tell you. Use what they tell you.*

## Introduction checklist and blacklist

When you have drafted your introduction, check whether you can tick all the items in Box 9.5.

### Box 9.5 Checklist for monitoring the quality of your introduction

- Do you demonstrate that you have understood the question?
- Do you summarize the key points of your answer?
- Do you show that you see beyond the superficial issues?
- Do you indicate the structure of your essay?
- Do you provide evidence of original thinking to set your essay above the crowd?
- Do you establish a clear and professional tone?
- Do you write fluently and without errors?

### *Conclusion to Chapter 9*

- The introduction is very important both as an aid to you in writing and as an aid to the reader.
- We've explained some basic rules for producing a safe introduction, but to go beyond this you need to apply your own wit and experience to develop a style of your own.
- You really should not even think about writing the middle bit of your essay until your introduction is sorted out. However, you will come back and revisit your introduction from time to time as the essay progresses, so you will have another chance to make it perfect before you hand it in.

10

# What goes in the middle of the essay?

## *Chapter summary*

What goes in the middle of your essay depends on what went at the front. The middle section should flow logically from the introduction and reflect the original plan. This is where you develop the 'plot' of the essay, identify and isolate the key points, and link these points into a coherent argument. You also need to provide evidence, case studies and illustrations for each point and for the links between points, and clear signposting to help the reader to navigate through the essay.

## What job does the middle bit of the essay have to do?

The middle bit of the essay has to do the following things.

- Make all the points that build up to create the structure of your essay.
- Present those points in a logical sequence with clear links between them.
- Provide evidence to justify each point.
- Provide examples to illustrate each point.
- Provide references to the sources of all information.
- Place all the information into the specific context of the essay title.
- Consider contradictory viewpoints or alternative interpretations.
- Provide a relevant response to the question or instruction in the title.
- Match up with what you wrote in the introduction.
- Lead logically into your conclusion.

## What goes in the middle?

It depends on what went at the front. If your essay is going well so far, then what has to go in the middle should be clear from the plan (Chapter 7) and the introduction (Chapter 9) that you have already written. Like most academic tasks, the hardest and most important part of essay writing is the preparation. Once the groundwork is done, finishing the job should be relatively simple. If, for reasons we probably don't want to know about, you are trying to write the middle bit of an essay that doesn't already have a plan and an introduction, stop it. It's not going to work. What goes in the middle *depends* on what went at the front.

Eric says: *The middle bit of your essay is just the extended version of the middle bit of your plan.*

The middle bit is basically the essay itself, shorn of the introduction that tells the reader what is coming and the conclusion that tells the reader what just happened. Forget all the build-up and the hype – this is where you have to roll up your shirt sleeves and knuckle down. This is where you tell your story. Start at the beginning and work your way through point by point. Your plan will tell you what to put in.

**You need to have your plan in front of you as you write the essay.**

## How do I develop the plot of the essay after the introduction?

When you devised your essay structure and put together your essay plan, you were constructing a story or an argument designed to consider alternative viewpoints and defend a particular conclusion. In Chapter 7, when we discussed structure, we worried about putting together logical arguments that led to sound, defensible conclusions. If you did that at the planning stage, all you need to do when it comes to writing the middle section is translate that logical structure or argument into a fully worded text. At this stage, although the prospect of committing words to paper can be daunting, you are not actually inventing anything new. The hard work was done at the planning stage and now you are just translating it. As you develop the plot of your essay, follow the plot that you worked out in your plan. If you need a reminder, look back at the introduction you just wrote.

**The introduction summarizes the plan and should remind you what comes next.**

When a reader finishes your introduction, it should be immediately apparent what to expect in the following paragraph. What comes immediately after the introduction is the first chapter in your story – your first point. That will lead into your second point, and so on. If you're stuck, re-read your introduction. It should be obvious what comes next.

Suppose your introduction said 'There are three main reasons for $x$.' The next paragraph would then logically begin with 'The first reason for $x$ is . . .'. The following sections then could be 'The second reason . . .', 'The third reason . . .', 'These three reasons relate in this way . . .', and finally 'Conclusion: there are three reasons, and they are interrelated.' It's simple, but it's clear. In the first instance this is what you must learn to do: produce a clear story in the middle of the essay that follows the plot that was arranged in your plan and summarized in your introduction. If you are in any doubt about how to proceed, keep it simple and just do what comes next in the plan. Look at Box 10.1 for some more examples of how an introduction leads naturally into the middle section of an essay.

## Box 10.1 Simple examples of how an introduction can lead naturally into the middle section of an essay

*Introduction 1:* There are three main reasons . . .

*Next paragraph:* The first main reason . . .

*Introduction 2:* This assessment is valid in polar regions but not in the tropics.

*Next paragraph:* In polar regions . . .

*Introduction 3:* Sustainability is a function of environment, political will and economics.

*Next paragraph:* Environment plays an important . . .

Middle sections generally have a straightforward bread-and-butter role to play in carrying forward the plot. You can add flourishes and plot twists if you want, but you should be careful not to *lose* the plot. The essay's plot is the story that you covered with your plan. If you look at Box 10.2 you will see some examples of simple essay plots. These should remind you of things we said about structures and plans in Chapter 7. That shouldn't surprise you. The middle of your essay is the fully written-out version of the plan. It is the manifestation of your structure.

## Box 10.2 Examples of simple essay plots

*Plot 1.* There are three parts to this answer. This is the first part, this is the second part and this is the third part. Here is a conclusion that ties the parts together and re-states the answer.

*Plot 2.* The answer to this question depends on whether it is answered with respect to arid or humid environments. In arid environments this is the answer. In humid environments this is the answer. Here is a conclusion that compares the possible answers.

*Plot 3.* The answer to this question is a complex explanation. Here is the chemical background to the explanation. Here is the core of the explanation. Here are some details that cover unusual situations. Here is a conclusion that summarizes the explanation.

## How do I identify and isolate my key points?

Both you and your reader need to be able to identify your key points. In order to write a clear essay that follows the points set out in your plan, you need to be able to recognize what your key points were and in what order they slotted together. That should be easy for you to do once you have got the plan sorted out. What is more difficult is to ensure that your reader also can distinguish your key points and see how they slot together. Your reader does not have the benefit of your beautifully drawn plan. To make sure that your reader can follow your plot point by point, you need to isolate and label your key points so that they stand out individually amid the pages of text. The easiest way to achieve this is through your paragraph structure.

We discussed the idea of paragraphs in Chapter 8 when we considered what makes a well-written essay. You should be familiar with the idea that each paragraph is responsible for dealing with an individual point from start to finish, including:

- a statement of the point itself;
- an explanation of how the point fits into your story;
- evidence, examples and references to back up the point.

**If each paragraph does its job and relates clearly to a point from your plan, your essay will stick to the plot and the reader will recognize each point as it comes along.**

If you run two paragraphs together, the reader may not recognize that you are making two separate points. If you split a point between two or three paragraphs, you need to take extra care to be sure that the reader recognizes that these are not three separate, half-completed points.

If you are not sticking to the simple point-per-paragraph structure, or if you think the reader may for any other reason need extra help (readers usually do), then you can provide that help through the internal structure of each paragraph. In the same way that essays benefit from having the key point spelled out at the beginning (Chapter 9), paragraphs also benefit from having the key point spotlighted at the start of the paragraph. Just as with the essay as a whole, a paragraph benefits from having the point written down right at the front in that it helps you to stay on target and shows the reader what is coming.

Eric says: *Each paragraph needs its own introduction, middle and conclusion.*

You can get a good idea of how clear and simple your essay will seem to a reader by skimming very quickly through your text, reading just the first line or two of each paragraph. Who knows, maybe that's what some tutors do when they mark essays? In the same way that you could find out what was in an essay by reading the introduction, you could find out what is in a paragraph by reading the first line. The sum of all the first lines should add up, more or less, to the introduction. This is a very mechanical and formulaic way of thinking about your essay, but as a way of starting to analyse your layout it can be helpful. As we said previously, you should aim to develop your style beyond this very basic level, but knowing the basics will at least keep you out of trouble. Look at Box 10.3 for an example of how an introduction can be made up from the first lines of the paragraphs of an essay, or vice versa.

### Box 10.3 A simplified example of how an introduction can be made up from the first lines of the paragraphs of an essay, or vice versa

*This example of a four-paragraph essay is, of course, rather short!*

*Title:* Why do some glaciers move faster than others?

*Introduction:*
Glaciers move at different speeds because their motion is controlled by a range of factors that can vary geographically and over time. The driving stress for motion is derived from the mass of ice under the influence of gravity. However, the response to the driving stress is controlled by ice rheology and basal friction, which are in turn controlled by environmental conditions such as temperature. The interplay of mass balance and flow dynamics determines the speed at which glaciers move.

*Paragraph 2:*
The driving stress for motion is derived from the mass of ice under the influence of gravity. Glen's flow law demonstrates that . . .

*Paragraph 3:*
A glacier's response to driving stress is controlled by ice rheology and basal friction, which are in turn controlled by environmental factors such as temperature. Under warm conditions, ice deforms . . .

*Paragraph 4 (Conclusion):*
The interplay of mass balance and flow dynamics determines the speed at which glaciers move. Where the driving stress is . . .

## How do I link the key points into an argument?

In your plan, points were linked with arrows and symbols, and relationships between sets of points were indicated by putting the points in little boxes or writing them in different colours. To achieve the same clarity in the text of your essay you need to use different techniques. Your fundamental tool, once again, is the paragraph. The opening and closing lines of each paragraph, which Eric referred to as the paragraph's introduction and conclusion, signpost the way into the paragraph at the top and the way out of the paragraph at the bottom. The most sensible way to link your paragraphs, and hence your key points, is by making sure that the signpost out of each paragraph points directly towards the signpost into the next. Look at Box 10.4 for an example of how paragraphs can link together.

### Box 10.4 An example of how paragraphs can be linked together

*Moving from discussion of point A to discussion of point B.*

'. . . [*coming out of the middle of paragraph 1*] . . . key point A, which is volumetrically the most significant point of the three that Bosch (1993) identified, is therefore only applicable in agricultural economies, and does not apply to industrial settings.

In sharp contrast to key point A, key point B applies only to industrial economies and, although it is globally less significant than point A, it is the dominant factor in the industrialized regions of Europe and North America . . . [*moving into the middle of paragraph 2*] . . .'

Sometimes, your essay structure does not require fluent linking between paragraphs. For example, if your essay is a simple list of three different things, set up with a statement to that effect in the introduction, then it can actually be an advantage to make sure that there is a very noticeable break between each item as it appears in the essay. You actually want to create a sense of a break. You want to put a rumble strip across the road so that the driver recognizes the shift into new territory. This is the kind of situation where you might consider using subheadings (see below).

## Can I use other techniques like bullet points to improve clarity?

Your aim is to write an essay that communicates clearly and effectively. If the use of a bullet-point list, a flow chart or some other device will help you to communicate, then you should consider it. However, as you have been instructed to write an essay, you must make sure that your use of these other techniques doesn't allow your essay to degenerate into a set of notes or a report. Remember, an essay is a very particular form of writing, and you don't want to drift too far into alternative forms. Different tutors may have different opinions about how far you should be allowed to go down the road away from straightforward sentences and paragraphs. If in doubt, don't go too far.

**You are going to be judged on your ability to write essays, not lists.**

You will get credit for judicious use of a variety of techniques of presenting material, but you will be penalized if the marker thinks that you are using these techniques because they are easier than putting together 'normal' text. It's good to break up text and introduce variety, but bad to break up text and lose the flow of the essay. Use 'alternative' techniques in moderation. If you can't remember what an 'essay' is supposed to involve, look back at Chapter 3.

## How bad is it to use subheadings?

One device that students often seem to want to use is the subheading. Subheadings are the ultimate big ugly signpost. You might notice that this book (like most) has subheadings! But it also has an index. The reason for both is that you don't expect someone to read a book from start to finish all the time. Often they just want a specific piece of information. The index and subheadings are the way they can find it. If you think the reader will need subheadings to follow your essay, your underlying structure and your use of standard signposting techniques are probably weak. There are exceptions to all our rules, but generally an essay is a single unit of work that it should be possible to organize coherently without subheadings.

Eric says: *If you can work out where to put subheadings in your essay, it must already be sufficiently easy to follow and so you don't need them. If you can't work out where you should put them, that's when you are in so much trouble that you could use them!*

One possible exception to this rule is where the essay title breaks the topic into clearly separated sub-topics. For example: 'List the main ingredients of (a) shampoo, (b) bubble bath and (c) pancakes.' In this situation it would be reasonable to break up your essay under three subheadings. Effectively you are being asked to write three separate essays. Box 10.5 gives some examples of titles that might give you the excuse to break the essay up into subheaded sections.

In other situations, where the title invites you to write on a single topic, you probably won't need subheadings. If you do want big ugly signposts messing up your essay, go ahead, but don't say we didn't warn you if your tutor doesn't like them. When an essay appears with subheadings it is hard for the tutor not to assume that the writer has trouble stringing together complex ideas in conventional text. Subheadings are more likely to suggest that you have weaknesses than to indicate that you have strengths.

## Box 10.5 Some examples of titles that might give you the excuse to break the essay up into subheaded sections

- With specific reference to three separate case studies, discuss the impact of town planning on urban morphology.
- Explain the impact of aeolian processes on landforms in (a) hot deserts and (b) periglacial environments.
- Review the history of the discipline of geography, discuss the contribution of geography to other branches of social and natural science, and identify the main challenges facing geography in the new millennium.
- Define three of the following terms: tombolo, desert, fractal, geography, society.

# How do I provide evidence for each point and for the links between them?

You have to provide evidence for everything that you say in an essay. Every assertion must be justified. Every statement needs to be backed up. Every claim needs to be supported. Every time you suggest a relationship you need to demonstrate that it is real. Every time you say that one point leads to the next you need to prove it. It's a pain, but that's the way it works. Get used to it. Get into the habit.

If you need motivation, imagine that your tutor will assume that you know nothing and that you are making everything up. Your tutor won't believe a word you say. Your tutor is so thick that you have to go through every single point step by step, with an example to show what you mean. Your tutor is such a sceptic that you have to provide a reference for every single piece of information to show you didn't just make it up. Writing an essay is like presenting a case in a court of law. You have to prove, demonstrate and support everything. You *must* show the evidence.

Evidence most often comes in one of the following forms:

- findings of previously published research;
- a real-world example of the thing you are talking about;
- a quotation from a published source;
- irrefutable logic.

Box 10.6 provides an example of each of these types of evidence.

## Box 10.6 Examples of different types of evidence being used to support a point

*1. Findings of previously published research:*

'The presence of moisture enhances the susceptibility of rocks to thermal weathering. This was demonstrated by laboratory experiments on eight different rock types reported by Bill and Ben (1990). They found that thermal weathering rates in moist environments were on average *x*% higher than in arid environments, with even the least susceptible lithologies experiencing a z% increase in weathering rates when moisture was introduced into the experimental environment.'

*2. A real-world example of the thing you are talking about:*

'In some remote areas ice from glaciers still provides a valuable natural resource. This is illustrated by rural entrepreneurs in Peru who collect ice from the margins of Andean glaciers, transport it by mule and truck to local markets, and sell it both as a refrigerant and as an ingredient in iced confectionery (Ben and Bill, 1991).'

*3. A quotation from a published source:*

'Controversy and rivalry in scientific research is very common. For example, Bill (1992, p. 35) wrote: "Ben's (1989) work on this topic is nonsense".'

*4. Logic:*

'Geographical research in the 21st century is likely to focus on issues that are relevant to environmental change. Research is partly driven by funding, funding follows politically important topics, and the changing environment is one of the most politically important topics in geography.'

You will notice that these types of evidence may overlap. For instance, in Box 10.6, example 3, a quotation from a published source, is also a real-world example of the thing we are talking about (like our example 2); and example 2, our real-world example, is drawn from previously published research. Also notice that sometimes the evidence you use to support one point then needs additional support of its own. Example 4, evidence from logic, really needs to be followed up with some evidence that our premises were justified. We can't just say that the environment is a politically important topic; we need to demonstrate that. A quotation from a politician, for example, could be useful here!

Eric says: *In your essay you need evidence to support your point, and evidence to support your evidence!*

The evidence that is most often used to support your evidence is the reference to published work. We will discuss references and how to use them in more detail in Chapter 12.

## How do I present my examples to make the most of them?

Examples are one of the things that really give your essay a chance to shine out above the dull essays of your fellow students. They are one of the easiest opportunities to make a big difference to your essay for relatively little intellectual input. Producing a unique and brilliant essay structure requires genius. Producing a unique and brilliant example just requires a bit of time in the library and a modicum of technical skill. We can give you that skill. However, getting hold of your good examples from the library is only the start.

**It is shocking how many essays waste their good examples by presenting them badly.**

The worst use of an example is the so-called 'e.g. Snowdon' case. This is where the student simply bungs in the name or label of their example with no detail and with no indication of how the example actually supports the point. Those are the two things that you need in order to make the most of your example. Consider the two cases in Box 10.7. These give two different ways of using the same example to support a point.

### Box 10.7 Two different ways of using the same example to support a point

*1. The wrong way to do it:*

'Many mountainous areas in the UK display evidence of glacial processes (e.g. Snowdon).'

*2. A much better approach:*

'Many mountainous areas in the UK display evidence of glacial processes. For example, Smith (1989) argues that the geomorphology of Snowdonia is controlled on a regional scale ($> 10$ km) by its geological characteristics but at a local scale ($< 1$ km) by features such as valleys, corries and peaks associated with glacial erosion and at a micro-scale ($< 0.1$ km) by features associated with both fluvial erosion (valleys) and glacial deposition (moraines).'

The good use of an example will impress the marker in a number of different ways. It shows that you:

- know how to make the most of your material, which is an important skill;
- are taking time and trouble to explain yourself clearly and completely;
- have some depth of knowledge about your example;
- understand the significance of the example for the essay.

That's a lot of brownie points for one little example, and it leaves the 'e.g. Snowdon' student trailing in your wake.

## Is my middle bit OK? A checklist

There are a few things still to add to the middle of your essay that we have not yet discussed because we cover them in later chapters. These include things like illustrations (Chapter 13) and references (Chapter 12). Before we move on to discuss those things, you should check that the core material of the middle bit of your essay is doing the key jobs that are required of it. When you have drafted your middle bit and you want to see if it is OK, remind yourself of the list of jobs that we said the middle bit of your essay was supposed to do. Does yours do these jobs?

- Make all the points that build up to create the structure of your essay.
- Present those points in a logical sequence with clear links between them.
- Provide evidence to justify each point.
- Provide examples to illustrate each point.
- Provide references to the sources of all information.
- Place all the information into the specific context of the essay title.
- Consider contradictory viewpoints or alternative interpretations.
- Provide a relevant response to the question or instruction in the title.
- Match up with what you wrote in the introduction.
- Lead logically into your conclusion.

### *Conclusion to Chapter 10*

- At this point you should have a fair idea of what sorts of things go into the middle of your essay, and what job the middle bit of the essay is supposed to do.
- You have by no means finished with that section yet, but we need to look ahead to what comes at the end of the essay before you can come back and polish up the middle section.
- We'll talk about how to put the finishing touches to the essay in Chapter 15, but before that we will next consider what comes after the middle bit.

11

# What goes at the end of the essay? The conclusion

## *Chapter summary*

**Your essay does not end when you make the final point of your story in the middle bit. You need a conclusion that helps the reader to appreciate the story you have told, and you can make your essay stand out from the crowd with a penultimate section that demonstrates your superior insight. The penultimate section could be a summative case study or a new angle on the question. The final conclusion needs to restate your key points, without being repetitive and with reference to subtleties that have been introduced during the middle bit of the essay. By the end of the essay you need to have reached the point that you declared yourself to be setting out for in the introduction, and your readers need to agree that they have reached that point with you.**

## How does your reader know when you have finished?

Will your readers know that they have come to the end of your essay and not just dropped the last page? It is important that they are in no doubt that you have finished. There are two ways that they will know. First, they will see that you have done everything that you set out to do in the essay. Second, you will tell them that you have finished by presenting a conclusion.

## Why not just stop when you finish the middle bit?

Usually it is good advice in academic writing to stop as soon as you have finished. Don't ramble on unnecessarily. This is true in essay writing, but the essay isn't finished at the end of the middle bit. We explained in the previous chapter that the middle bit contains all the core material that tells the story of your essay; but your essay is more than just the basic story. In the same way that we put an introduction at the front of the story to set the scene, signpost the route, and point out the highlights in advance, we also need some closing sections.

Without a closing, your essay will dribble to a stop with the reader looking at the wet paint on the last piece of evidence that you gave them for your last key point. What you want is for it to end with a fanfare and applause, with the reader holding a gift-wrapped package containing a framed print of your complete picture. Without a closing, your essay will leave the reader just at the point where their train pulls into the last station on their journey. With a good closing you can welcome them to their destination, help them with their luggage (or intellectual baggage!) and check them into their hotel. Incidentally, we don't recommend that you use analogies like this in your essays.

## What goes at the end of the essay?

Everybody knows that you need a conclusion at the end of an essay. However, there's more to it than that. Certainly you will need a conclusion at the end, and if you are aiming for a basic essay with no frills then you can move straight on from the middle section into your closing statement. If you are aiming for an essay that does more than the bare minimum, then there is a useful additional section that you can insert into your essay in between the middle bit and the conclusion. Let's call it the penultimate section. It's a sort of a pre-conclusion.

## What is the penultimate section?

The penultimate section comes after you have completed the story that occupies the middle bit, and before you move into the summing up that will go into the conclusion. You could miss out this penultimate section completely, and many students do so, but if you include it you can produce a much more impressive essay.

**The purpose of the penultimate section is to do something extra that sets your essay above the crowd.**

It needs to be something useful and relevant, not just a superfluous add-on, but something that not everyone will think to include. Since originality is one of the great strengths of a good penultimate section, you should try to think of your own ideas for what to include, but useful approaches include:

- the summative case study;
- the new angle;
- the wider context.

### Penultimate section, type 1: the summative case study

Throughout the middle bit of your essay you should have been using examples and case studies to demonstrate and illustrate the points that you have been making. However, not everybody remembers that their essay, in addition to making a whole series of individual points, adds up to one big point. The overall message or point of your essay needs an example as much as the individual points within each paragraph do. The summative case study is a special example that relates not just to an individual part of the argument, but to the whole assemblage of points that are going to come together to make your final conclusion. That's why the summative case study makes a good penultimate section: it makes a good structural link between the middle bit and the conclusion.

**The problem with using a summative case study is that you really have to find a good one.**

If you choose a case study that doesn't cover the whole breadth of your answer, the reader will think you've just stuck in yet another little example like the ones you've already used half a dozen times earlier in the essay. Not only will this fail

## Box 11.1 A simple example of how a summative case study can bring together all the different parts of your story in a single example

*Title:*

To what extent has the success of local fishing cooperatives in region 'x' been limited by the size of their markets?

*Key points:*

1. Close correlation between economic success and market size. *Examples:* a failing cooperative with a small market; a successful cooperative with a large market.

2. Poor correlation between economic success and fish production, cooperative structure or marketing strategy. *Examples:* cooperatives with similar production, structure or strategy meeting different fates.

*Summative case study:*

Cooperative 'y': changing market over time leads to changing fortune although other factors (production etc.) remained constant. Unsuccessful until closure of neighbouring cooperative led to increase in market potential. Rate of growth of profit of cooperative 'y' correlates with progressive increase in market upon successive closure of three neighbouring cooperatives.

*Tutor's comment:*

The summative case study illustrates both of the essay's key points, which focus on the difference between cooperatives with small and large markets, by combining them in the example of a particular cooperative that experienced a change in market area over time.

to do the job you need, it will also have the negative effect of distracting readers from the overall synthesis that should be coming together in their minds at this point. A good summative case study has to be an example of why your whole essay was right; of how your whole argument hangs together; of how all the different parts of your story apply in a single instance. Box 11.1 gives a simple example of a summative case study that works in this way.

## Penultimate section, type 2: the new angle

Throughout your essay you will have considered alternative points of view and presented balanced assessments of contradictory sets of evidence. However, your essay will almost certainly have been written within the single framework of your own understanding of the question, within the paradigm of your particular approach to the answer. There's nothing wrong with that, but think about the assumptions that you have made in interpreting the question and setting up your answer. For example, if the question was about landslides, did you assume you were only supposed to talk about Earth? If the question asked about the future of local government planning strategies, did you assume that local government would continue to exist? Almost certainly, these fundamental assumptions that you made in setting up your answer were sound, logical, and absolutely the right assumptions to put at the core of your essay. Think for a minute about whether your essay would have had a fundamentally different answer if you had made a different assumption. For example, would landslides work differently on another planet with different gravity? Would planning strategies be different if local government were replaced by centralized control? If so, you have the opportunity to produce a 'new angle' penultimate section.

### Box 11.2 Some examples of plausible 'new angles' for a variety of essay topics

*Topic 1:* Hazards associated with volcanoes.

*Core material:* Lava, ash, lahars, gas etc.

*New angle paragraph:* Additional hazard caused by efforts to monitor volcano.

*Topic 2:* What controls glacier velocity?

*Core material:* Ice deformation, basal friction, water etc.

*New angle paragraph:* Since gravity is key, glaciers on other planets different.

*Topic 3:* Future of town planning in UK.

*Core material:* Environmental, social, political and economic goals and constraints.

*New angle paragraph:* What if a surprise new political party came to power?

The trick with the new angle is to find an angle, an alternative scenario, that is sufficiently novel that it did not need to be considered as a matter of course in the core of the essay, but sufficiently plausible that it is not completely irrelevant or just plain silly. Box 11.2 gives some examples of plausible 'new angles' for a variety of essay topics. Be careful not to go beyond the realms of plausibility. Although extraterrestrial environments are always good ground for inspiration, encounters with extraterrestrials rarely are.

### Penultimate section, type 3: the wider context

Your essay title will have limited you to write about a specific topic and, because you have followed our guidelines, you will not have strayed from the topic and rambled on about related, but not strictly relevant, material that you happened to know so you thought you'd stick in anyway. However, there is a place for such material, and this may be it. By now you will have convinced your tutor you know all about topic *x*. Now is your chance to gain extra marks by putting in additional material that would have distressed your tutor early on, but will now show you appreciate where the title fits into the wider scheme of things. So, if your essay were about anthropogenic causes of climate change, now would be a chance to set the scale of those causes in a wider context of natural processes. Is human activity in the future going to cause climate to change on the scale we have seen, as a result of natural processes, in the last 10 000 years? Addressing an issue like that in a short penultimate paragraph shows you can't just do what's asked of you but can give that bit extra.

## How do I make sure the penultimate section really works?

The penultimate section is an opportunity to show the reader that you are a little bit more insightful, original, or knowledgeable than might otherwise have been evident. If you have written a penultimate section and you wonder if it's any good, or if you are trying to think what to put in, think about this list of the hidden messages that your penultimate section could be conveying to the reader:

- I bet no one else thought of this.
- I have seen more deeply into the question than you expected.
- Look how laterally I can think.

Don't forget, when writing this section, that it needs to be signposted just as well as all the other sections. In fact, since not all students will include this section, signposting is probably even more important here than anywhere else. Your tutor may not be ready for this section. If your 'alternative' viewpoint is mistaken for part of your core argument you could give the wrong impression.

Signpost at the top of the section that this is an all-embracing example, an alternative angle on the question, or whatever it is that you have tried to achieve. Signpost at the bottom of the section that you have finished that topic and that you are now moving into the formal conclusion, where the alternative and additional gives way to the core and essential. That's exactly what we are doing now (signpost!).

## What job does the actual conclusion have to do?

The conclusion has a number of jobs to do.

- Sum up the essay's final position (your answer) in response to the title.
- Remind the reader of the main points that led you to that position.
- Point out any problems, controversy or uncertainty in your position.
- Judge the significance of your position.
- Show that what you have done is completely relevant to the title.
- Match up with what you said in the introduction.
- Leave the reader with a positive final impression.
- Tell the reader you've finished!

Students sometimes ask whether, since the conclusion is summing up the answer to the question, it is just a repeat showing of the introduction.

**Certainly there should be common ground between your introduction and your conclusion, but there are fundamental differences.**

The introduction identified the overall answer and the types of evidence that you would go through in presenting it. The conclusion can give a much more specific answer including all the subtleties and points of uncertainty that have been drawn from the evidence in the core of the essay. If the introduction is like the trailer to a film that lets the uninformed spectator know roughly what to expect, the conclusion is more like a review that discusses with the experienced cinema-goer exactly what was screened.

Summing up the essay's final position and reminding the reader of the main points that led you to that position is the 'thus it can be seen that . . .' part of the essay. However, people who start their conclusions with that phrase are the same sort of people who begin their introduction with 'the dictionary definition . . .'. Students who write like that are unlikely to rise above the lowest level of lower second class grades. Any student who is good enough to get a better grade than that should know better than to write in hackneyed clichés that were old when Noah was a boy. Instead of starting your conclusion with 'Thus it can be seen that . . .', just start it with whatever would have come next. Instead of writing 'Thus it can be seen that there are three main consequences of alien invasion', just write 'There are three main consequences of alien invasion'. If the bare phrasing of a simple statement of your main point seems inadequate without the frill of an introductory clause, it is probably because your answer is itself too bare. Is your conclusion really just that there are three main consequences, or have you in fact shown more than that? The opening of your conclusion would sound much better if, instead of writing 'There are three main consequences' you could write 'Although three main consequences have been identified by recent research, most authorities agree that consequence (A) far outweighs the other two in its international significance.' Rather than just repeating your original answer, you are repeating it with some added nuance derived from the evidence presented in your essay. You are ending the essay by showing that you've actually progressed from your introduction and got somewhere.

## An end-of-essay checklist: make sure that yours does all these things

When you have written the final sections of your essay, read through and check that you can answer 'yes' to all these questions:

- Have you provided an interesting penultimate section that makes your essay different from those of other students?
- Have you summed up the essay's final position (your answer) in response to the title?
- Have you reminded the reader of the main points that led you to that position?
- Have you pointed out any problems, controversy or uncertainty in your conclusion?
- Have you assessed the broader significance of your answer?
- Is what you have written completely relevant to the title?
- Does what you've written match up with what you said in the introduction?

Eric says: *There should be nothing in the conclusion that comes as a surprise to the reader.*

## Conclusion to Chapter 11

- The conclusion is a difficult section that has a lot of ground to cover.
- This is your last chance to impress the reader and to make sure that your point has come through loud and clear.
- It's also your last chance to show that you understand the subtleties and controversy surrounding the issues that you have discussed, but that you can nevertheless see through the clutter of detail to produce an insightful response to the title.
- You should not introduce new material in the final paragraph, but you may use your penultimate paragraph to consider the key issues in your essay from a new point of view.

# 12 How do I use references and write a reference list?

## Chapter summary

References are your method of mentioning previously published work and acknowledging the sources of information or ideas that are not your own. You include references to show that you aren't inventing evidence for your argument, to demonstrate the breadth of your involvement with the literature, to lend support to the points you make in your essay, and to allow readers to trace the source materials that you have used. There are different systems for presenting references, but they all include a short reference in the text and extended bibliographic details in a separate list. You won't usually be expected to present a reference list at the end of an exam essay, but for coursework essays it is important to understand references and to use them properly. If you don't provide references you will look ignorant and you may be suspected of plagiarism.

## What is a reference?

A reference is the way that your essay points to a previously published piece of work or to the source of any information that you use in your essay. References can also be called citations, and when you make a reference in your essay you are citing what somebody else has said or done previously. A reference is like a note to your readers telling them about something in a book or a journal or some other source. In one sense you are *referring* to the work in that you are mentioning it. In another sense you are *referring* your readers to the work in that you are pointing them towards it.

Eric says: *A reference is your way of mentioning something that's been published before, or giving credit to somebody when you use their material.*

## Why do I have to include references within my essay?

You need to have references:

- to show where you obtained specific pieces of information;
- to indicate the sources of ideas or words that are not originally your own;
- to identify where quotations come from;
- to list previously published work that is relevant to your essay;
- to provide evidence for points in your argument;
- to demonstrate your familiarity with previous research in your field;
- to identify where readers can trace original publications that you mention;
- to avoid charges of plagiarism when you use other people's ideas;
- to show that you're not just making stuff up;
- usually, to meet the requirements of your assignment.

Just imagine if your essay didn't do any of the things in that list. What kind of grade would you expect to get? That's what would happen if you didn't include references. That's why you have to include references in your essays.

## Where in the essay should I put references?

You should put a reference wherever you use facts, figures, examples, opinions, words or diagrams from another source, and wherever you want to mention previous work that is relevant to what you are writing about. How you should insert the reference into the text will be explained in the next section, but Box 12.1 shows you some examples of situations where a reference could be necessary.

### Box 12.1 Some situations where a reference could be necessary

- The number of chickens in the area in 1936 was 258 000 (Smith, 1990).
- A great deal of research has been done on this topic (e.g. Smith, 1990; Jones, 1991).
- Eggs are important (Smith, 1990).
- According to Smith (1990), many researchers believe that eggs are important.
- Smith (1990, p.17) wrote that 'eggs are important'.
- Exaggeration is dangerous, as shown by the errors discovered in Smith's (1990) conclusions by Jones (1992), but the importance of eggs is demonstrated by the number of omelettes cooked each year (Shelby, 1993).

  Figure 2 is taken from Jones (1992, Figure 5).

Whenever you write down a piece of information that you obtained from a book or an article, you must provide a reference to the book or article. Whenever you report an opinion held by somebody else, you must provide a reference to where you saw that opinion expressed. Whenever you use a quotation, you must provide a reference to the place it came from. Whenever you use an idea that you picked up from somebody else's work, you must provide a reference to that work. Whenever you include anything in your essay that is not a completely independent product of your own imagination, insight or original research, you must provide a reference indicating the source.

Eric says: *If in doubt, reference it.*

If you don't cite references where you should do, two serious consequences could arise. First, your essay will be weaker than it would be otherwise. Second, you could be suspected of plagiarism. Plagiarism is basically nicking an idea or some information from somebody else and pretending that it is your own. It's academic cheating, and you really don't want to be suspected of it. We will say more about plagiarism later in this chapter and in Chapter 21, but for now suffice to say that you must reference your sources.

## How do I include references within my essay?

You cite references by mentioning the relevant source briefly at the appropriate point in the text of the essay, and including full bibliographic details of the source separately at some location outside the main body of the text. There are several different styles or conventions by which you can do this. You need to choose the one that is most appropriate to the type of sources that you are citing, or the one that your tutor has instructed you to use!

The two most common ways of presenting references in geography are the 'Harvard system' and the 'Vancouver system'. The Harvard system is based on putting the author's surname and the year of publication in the text, and putting an alphabetical list at the end of the essay that gives all the bibliographic details of all the references cited in the text. The Vancouver system is based on putting a little number in the text, like this (1), to indicate that there is a reference relevant to this point, and putting the bibliographic details either in a footnote at the bottom of the page or in a numerical list at the end of the essay. The Harvard system is very convenient when your sources are mainly published books and articles with clear authorship. The Vancouver system has advantages where many of your sources have no clear authorship, such as birth certificates or parish records.

You will encounter many examples of both of these types of reference system as you explore the literature in geography and related disciplines. The Harvard system is by far the most commonly recommended for use by geography students and will generally be the most appropriate for you to adopt in your essays. Your institution, course or tutor may want you to use a particular style in your essays, or may ask for different pieces of work to use different systems, so be sure to check your instructions for the specific requirements of your assignment. If you are given no specific direction, then the Harvard system is usually the most appropriate one for you to follow, and that is the system that we will focus on here.

Whatever system you use, there are two key points to bear in mind:

1. Be consistent throughout your essay.
2. Provide enough bibliographic information for readers to trace the original item that you are describing, should they wish to do so.

Eric says: *If your reference list is done properly you will have provided enough information for your readers to order a copy of each item from the library.*

# The Harvard system

There are two elements to a reference in the Harvard system: a short reference in the text of your essay and the full bibliographic details in the list at the end of the essay. We will deal with each in turn.

Within the text of your essay you need only give the surname(s) of the author(s) of the item you are referring to, plus the date of publication of the item. If the item has more than two authors, name only the first author and add '*et al.*' to indicate that there were others (*et al.* is an abbreviation of the Latin *et alia*, meaning 'and others': notice that there is no full stop after the '*et*' as it is not abbreviated). If the same author has more than one publication with the same date, so that two items look identical when you just put name and date in the text, then differentiate between them by adding letters (a, b, etc.) after the date to distinguish them. If, and only if, you are quoting directly from the source text, you should also include the page number.

You can write the reference as an integral part of a sentence, or you can add it to the sentence as a supplementary item, perhaps as part of a short list of sources. Boxes 12.1 and 12.2 show several examples of how you can include references in your text. Look in any research journal or advanced text book for more examples.

## Box 12.2 Examples of how you can include references in your text with the Harvard system

- Field descriptions by Smith (1991) demonstrated this clearly.
- This has also been demonstrated by laboratory experiments (Smith, 1992).
- Smith's (1991) results contradicted those of Jones (1990).
- Contradictory results have been produced (Jones, 1990; Smith, 1991).
- Smith (1993, p.17) wrote 'My experiment was flawed.'
- The admission 'My experiment was flawed' (Smith, 1993, p.17) is relevant.
- Jones *et al.* (1993a; 1993b) condemned Smith's (1991; 1992; 1993) research.
- Relevant contributions include Smith (1991) and Jones *et al.* (1993a; 1993b).

From the examples in Box 12.2 you will notice the following rules:

- In references, only use authors' surnames, never first names or initials.
- Use '*et al.*' rather than listing all authors of multi-author works.
- If the same authors have more than one item per year mentioned in your essay, label them with letters.
- In a list of references within the text, separate references with semi-colons.
- In a list of references within the text, place items chronologically.
- Only include a page number if you are quoting directly from the source.

At the end of the essay you need to include a 'reference list' that contains all the details of each of the references that you cited in the text. The reference list should be ordered alphabetically by the surname of the lead author. Different types of source material (books, journals, newspapers) require slightly different presentation in the reference list, but the broad principles are always the same for all items. You need to supply:

- the name(s) of the author(s);
- the date of publication;
- full bibliographic details of the item.

When we say you need to present full details of the item, we mean that you need to present enough information that your readers could go straight to the item in the library, or go right to the bookshop with a fully completed order form. Box 12.3 shows you how to do this for some different types of source material. For each example we show you the standard format and give you an illustration. We don't cover every possibility, but this should be enough to show you how the system works.

**Box 12.3 How to write entries in your reference list for some different types of source material**

*For a journal article:*
Author(s), date, article title, journal, volume, issue, pages.

Smith, A. (1990) My article. *Journal of Important Articles* Vol. 24(12) pp. 12–56.

*For an authored book:*
Author(s), date, book title, publisher, place of publication.

Smith, A. (1990) *My book (3rd edn)* Big Publishing Co., London.

*For an edited volume:*
Editor(s), date, book title, edition, publisher, place of publication.

Smith, A. (ed.) (1990) *My edited book.* Big Publishing Co., London.

*For a single paper or chapter in an edited volume:*
Author(s), date, chapter title, (In:), Editor(s), book title, edition, publisher, place of publication, page numbers of chapter.

Jones, A. (1990) My chapter. In: Smith, A. (ed.) *My edited book.* Big Publishing Co., London. pp. 126–139.

*For a newspaper article:*
Author(s), date, article title, newspaper, date, pages.

Smith, A. (1990) My article. *The Paper News* March 11, p. 12.

*For an electronic-journal:*
Smith, A. (1990) My article. *Journal of Stuff* (Electronic) Vol. 24(12).

http://www.scienceavailable/r67890-t67

Accessed 21 October 2001.

*For an Internet site:*
Author, date of creation, title (online), URL, date accessed.

Smith, A. (1994) *My web page* (online). http://www.alfsmith.co.jp/stuff.html

Accessed 17 May 2003.

## What is the difference between a reference list and a bibliography?

The difference between a reference list and a bibliography is that a reference list only covers items that you explicitly cite in your essay whereas a bibliography can cover any sources relevant to your topic, whether or not you actually cite them in your essay.

A reference list is a list of all the references that you have cited in your text. It doesn't matter whether you actually looked at these sources yourself or whether you found out about them indirectly from another source. All that matters is that you cited them in your essay.

**If it's in the essay it has to be in the reference list, if it is not mentioned in the essay then it must not feature in the reference list.**

If you put things in your reference list that were not mentioned in the text of your essay, your tutor will probably think that you are pretending to have a more well-informed essay than you really do. This could be counted as a form of cheating and will probably be dealt with severely. If you cite items in your text that do not feature in your reference list, (a) your readers will be unable to track down these items if they should want to see them for themselves and (b) your tutor will think you are careless or incompetent.

A bibliography is a list of items that are relevant to the topic of your essay. It might include items that you actually mention in the text, items that you have used yourself but didn't actually mention, and items that you have never used but know from other sources to be relevant. A bibliography can contain anything that seems interesting. You will not usually be required to have a bibliography at the end of your essay.

If you have a reference list at the end of your essay, which you should do, make sure that you don't call it a bibliography, which would be confusing and will make you look stupid. If you find yourself writing a bibliography, check whether that's really what you are supposed to be doing. Probably, if this is a regular essay, you should really be producing a reference list. Producing a bibliography is a whole separate kind of assignment, and we are not dealing with that here.

## Are some types of reference better than others?

References that do the job they are being used for are better than references that don't. It's all a question of what's appropriate to your particular situation. If you are using a reference to lend authority to your evidence, you should use a reference that people are likely to trust. For example, if you are presenting evidence for trends in economic performance, a reference to an academic paper published in a refereed economics journal will carry more weight than a reference to a story in a tabloid daily newspaper. By contrast, if your reference is intended to provide an example of what kind of articles are published in tabloid dailies, then a reference to such an article would be appropriate.

Generally, however, some types of reference will impress your tutor more than others (Box 12.4).

## Box 12.4 Good and bad types of reference

*Good:*

- articles from academic journals (especially classics and new research);
- papers from professional and academic conferences;
- original data sources (census, archives, historical documents, etc.).

*Bad:*

- lecture notes;
- A-level notes and textbooks;
- personal communication from old school teacher;
- newspaper stories, unless these count as original sources for your topic;
- TV documentaries;
- most web sites.

The reason that academic papers score so much more highly than newspapers, web sites, TV and personal communications is that the information they contain has already been through a review process. It has been checked. This certainly does not mean that everything you read in an academic paper must be true or right, but it does mean that it has been more carefully checked than most newspaper stories. When you use any reference, think hard about how reliable your source is. Your tutor will. Take care not to base conclusions on unreliable sources.

# Do I have to give references in exam essays?

In exams you would not usually be expected to be able to cite the full bibliographic details of your sources of information, and so you will not be required to present a reference list. However, you will get credit for showing that you are familiar with the research literature in your topic, and you can do that by including relevant references in your essay. For example, a student who writes in an exam that 'subglacial observations have indicated that regelation occurs at the glacier bed' will not gain as much credit as one who writes that 'subglacial observations have indicated that regelation occurs at the glacier bed (Kamb and LaChapelle, 1964)'. The student who writes that 'subglacial observations have indicated that regelation occurs at the glacier bed (Agassiz, 1864)' will be penalized for using an incorrect reference, because it wasn't Agassiz who did this work.

**Only use a reference if you are sure it is correct.**

If you use the wrong one your tutor will think you are either ignorant or trying to cheat by pretending to know stuff that you don't. References to the course textbook are generally not very impressive: course textbooks rarely say anything that hasn't been said before, and a reference to the original research is usually more useful.

Eric says: *Do you get bonus marks if you cite your tutor's papers?*

One way to get some real bonus marks is to include a reference to a newly published article in a major journal that has appeared between the end of course teaching and the date of the exam. This shows that you have kept up to date with the ever-advancing research frontier, and that you've put in some good research and revision time.

## What is plagiarism?

Plagiarism is a serious form of cheating that involves copying other people's stuff. This might be copying Eric's essay, lifting chunks of text from the internet, or using an idea from a book without citing the source. We mention it here because you might be suspected of plagiarism if you do not correctly cite references to the sources that you have used in putting together your essay. If ideas are your own, or if information is based on your personal observation, then of course there is no source to cite. If your ideas and information are taken from books or articles, as will usually be the case with your geography essays, then you have to declare the fact by giving the reference. If you lift material from the internet you must present it as material lifted from the internet and give the source. If you don't give a reference you imply that the material is your own. If you do that when it isn't, that's plagiarism. Incidentally, don't think your tutor won't notice.

**Material that is not your own usually stands out very clearly to an experienced reader.**

Suspicious tutors type sections of student essays into search engines such as Google to see whether the text was lifted from the internet. If you cheat, there's a good chance you'll be caught. If you are not cheating, then make sure nobody can mistakenly think that you are. Cite references.

## *Conclusion to Chapter 12*

- Make sure you understand what references are and how you are supposed to use them.
- Make sure you understand the difference between a reference list and a bibliography.
- Make sure you use plenty of references in your essay, but make sure you only use appropriate references in appropriate places.
- Don't use material without giving a reference to the source, or you might be accused of plagiarism.

# 13 How do I use diagrams and other illustrations?

## *Chapter summary*

Illustrations such as maps, graphs and line drawings are collectively known as figures. Figures and tables can be a big help in an essay, but only if they are properly used and well presented. You must refer to each figure within your text, and figures must be presented with a caption that explains what they show. You must remember to cite the sources of the material in your figures, and you should only use material that is strictly relevant to the points that you are making. For each figure that you plan to include in your essay, you can use our checklist to make sure it's up to scratch. If your figures are good, they can make a good first impression even before the reader starts to read your essay.

## Should I use illustrations in my essay?

There's an old saying that a picture is worth a thousand words. However, you need to be careful, because a bad picture is worth a thousand bad words. A relevant, well-placed, well-presented figure can help you to explain a point and can add a huge amount of additional information that would be virtually impossible to include as text. For example, a map can show the location of a site or the global distribution of a phenomenon that could take hundreds or thousands of words to describe, and a photograph can enable a reader to appreciate a scene in more detail than you would easily be able to provide in words.

The main benefits of illustrations are that they:

- allow you to convey additional material that doesn't fit in the text;
- allow you to convey information clearly without using lots of words;
- can be used as structural pointers to separate sections of text;
- can be used to brighten up the appearance of your essay.

However, there are dangers in using illustrations. First, a poorly drawn, incorrect, misleading or inappropriate figure can do a lot of damage to your essay. Second, using illustrations actually gives you extra work because every figure needs to be appropriately introduced in the text and accompanied by a clear and detailed caption. Third, illustrations can actually obscure your point if they provide too much information, and you need to make sure that the reader is specifically directed to what is relevant. For example, if you include a photograph to show what a landscape looks like, you need to make sure that the reader understands whether the important point about it is the cliff in the foreground, the mountain in the background or the flock of geese in the middle distance. If you provide a photograph of just the cliff, you still need to point out whether it is the gradient, the vegetation or the colour of the weathered rock that matters. If you point out the colour of the weathered rock, you need to make sure that the reader knows which bits of the rock in the picture are the weathered bits. Making a diagram work for you is hard work.

## What kind of illustrations can I use?

Unless you have received instructions to the contrary, you can use any type of illustration that you wish. The most common types used in geography essays are:

- maps
- line drawings
- flow diagrams
- graphs and charts
- data tables
- photographs
- field sketches
- topographic cross-sections.

Different types of figure will be appropriate to different situations in your text, and your main concern should be to use *appropriate* illustrations. To present numerical data you should use a table or a chart. If different parts of the data show some correlation, such as a trend in the value of a variable over time, then a graph may be most appropriate. If you want to illustrate location or spatial distribution, then you need a map. If want to describe a landscape, then a photograph may help. If you are trying to describe complex systems or relationships, then perhaps a flow chart is what you need. Reading widely in previously published literature is a good way of getting ideas for the different types of figure that you might use.

**To get ideas that might be a little different from those of other students writing the same essay, look in literature outside geography.**

Astronomers, musicologists or electrical engineers might use some kind of illustration that does just the job you need. Don't be afraid to use novel approaches if you think they will work for your essay.

## When should I use an illustration?

You should use an illustration only when it will improve your essay, in other words when you are making a point that would benefit from the additional support that the illustration can provide. Don't just bung in a picture because you happen to have one. Use your illustrations at specific locations to assist your text.

You should use an illustration when you need something that will do one of the jobs that we suggested illustrations were good for in the first section of this chapter. Use illustrations primarily in places where they help to make your essay clearer or more informative. Don't use an illustration if it does not contribute directly to the relevant content of your essay. You could be penalized for including irrelevant or distracting material. Box 13.1 illustrates some situations where you could usefully include a figure to support a point.

### Box 13.1 Some situations where you could use illustrations to support a point

- When describing spatial distributions, use a map.
- When describing numerical distributions, use a graph.
- When describing a landscape, use an annotated photograph or field sketch.
- When listing large amounts of data, use a table.
- When describing a slope profile, use a cross-section.
- When describing the internal structure of a landform, use a line drawing.
- When describing an organisation or system, use a flow chart.

## How should I number my illustrations?

Each illustration that you include needs to be given a number so that you can identify which one you are talking about when you mention it in your text. It is normal to distinguish between tables and other types of illustration. Tables are referred to as Table 1, Table 2, etc. and all other illustrations as Figure 1, Figure 2, etc. If you prefer, you can also distinguish between different types of figures. For example, you could label photographs separately from the other figures. This convention stems partly from publishing practice where the technology of printing requires drawings, tables and photographs to be handled separately, and photographs sometimes do appear all together in one part of the book. Figures 1, 2, etc. would be maps and drawings; Plates 1, 2, etc. would be photographs assembled in one part of the book; and Tables 1, 2, etc. would be items set up and printed within the text.

If you are putting all your illustrations into the essay at the appropriate point in the text where you discuss them, then there is no reason for you to have separate systems for each type. You can label them in the order that they first appear in your text and it will be easy for the reader to find each figure when you mention it. However, if you are going to put all your photographs on separate sheets at the end of the essay, the order in which figures are introduced in the text, the order of their numbers, and the order in which they appear on the page will not correspond. In that case it is better to have a separate numbering system for the photos.

If you have complex figures that include more than one part, you can label the parts (a), (b), etc. For example, if Figure 1 is a set of four graphs printed one above the other, and you want in your text to refer to the second graph down the column, the easiest way to do it would be to label the four graphs (a), (b), (c) and (d) on the figure, and then in your text refer the reader to Figure 1(b). The caption would then take the form: 'Figure 1: Graphs of (a) this, (b) that, (c) the other, and (d) something else.'

If in doubt, just remember:

- Give every illustration a different number.
- Keep your numbering system as simple as possible.
- Number the figures in the order in which they first feature in the text.

## How do I write figure captions?

The figure caption is the little bit of text that you put with the figure to let the reader know what it is. The caption needs to be a freestanding unit of text that would make sense even to somebody who was not reading the text of the essay.

The caption needs to:

- say what the figure is;
- point out the key features of the figure.

Box 13.2 gives some examples of figure captions to show how this can work, as well as a couple of bad examples to show what you should avoid.

### Box 13.2 Examples of figure captions, adequate and otherwise

*Adequate examples:*

Figure 1: A map of central Ecuador showing the locations of the active volcanoes mentioned in the text and of the major cities that would be at risk in the event of major eruption (based on data in Smith, 1990).

Figure 2: Particle size distribution of glacial till from Snowdonia, showing the bi-modal pattern typical of glacial deposits. Source: Jones, 1999.

*Bad examples:*

Figure 1: A map.

Figure 2. Particle size distribution, Snowdonia.

Figure 3.

Figure 4: There are several active volcanoes in Ecuador that would threaten major cities in the event of an eruption.

## Should I mention my figures in my text?

You must mention each of your figures within your text. Readers making their way through your essay may not stop to look at the pictures unless you tell them to. Equally, they may stop at the wrong point and fail to see the relevance of your illustration.

**Each of your figures should relate to some specific point that you make in your essay, so you should mention the figure when you make that point.**

For example, if you were making the point that there are cities in Ecuador that could be at risk from volcanic eruptions, it would be sensible to say 'Hey, here's a map that shows what I mean' and to put in a map showing how close the cities are to the active volcanoes. Of course, you wouldn't write 'Hey, here's a map'; you'd write: 'Figure 1 shows the locations of active volcanoes in central Ecuador and of the cities that could be at risk from major eruptions.'

In some cases, mentioning the figure in the text can replace a whole section of text. For example, if Figure 2 is a data table, you don't need to go through all the data in your text. Instead of describing in words the quarterly ticket sales of three different airlines for each of five different holiday destinations, you can in your text simply refer to your data table: 'Figure 2 shows the quarterly ticket sales of three different airlines for each of five different holiday destinations.' This could save you several hundred words, and also make the data much more accessible to the reader than it would be as a text description.

## Are there special rules for certain types of figure?

Different types of figure do have certain specific requirements. For example, a map should always have a scale and an orientation. A graph should always have the axes labelled. A topographic cross-section should always include a vertical as well as a horizontal scale. Mostly these rules are just common sense as long as you remember what you are trying to achieve. Your goals are to ensure that the readers know exactly what they are looking at and that there is no possibility of misunderstanding or misinterpretation. There must be nothing ambiguous. Take a look at the checklist at the end of this chapter for some ideas about the things you need to look out for, and read the section immediately below this one for some ideas about what makes a well-presented figure.

## What makes a well-presented figure?

When your tutor marks your essay, you can expect feedback on the quality of your figures. You should be hoping for comments such as 'well drawn and informative', 'clearly presented and effective' and 'professionally executed'. There are some simple guidelines to follow that will earn you these kinds of comments.

If you are drawing a figure yourself, you need to make sure it is clear and neat. Don't load the picture with unnecessary detail and don't expect a few scribbled lines of biro or pencil to create a good impression. For the most professional effect you should draw your figure using a computer graphics package. Most undergraduate geography courses include some sort of training in basic drawing packages and computer cartography. Make sure you have read the instructions and practised the exercises that you have been given in your course. If you have not already received some training, there are whole textbooks written about design and execution of technical drawings and illustrations, and if you want more advice on this aspect of producing your essay you should refer to this literature.

There are some very basic guidelines that you could bear in mind. For example, if you are drafting figures by hand, you should draw in black ink at a scale larger than the intended final version, and then reduce the illustration on the photocopier or scan it into your word-processing package at a reduced scale. The scale reduction will tend to hide any slight wobbles in your drawing, variations in line thickness, or uneven lettering. Where possible add labels and lettering mechanically on the computer or use dry-transfer lettering or stencils to produce neat text.

**Hand-written labels rarely look good.**

If you are using a figure that you have found in another source such as a book, a journal or a web page, there are additional things to bear in mind. First, you must reference the source of your material. If you lift a figure directly from Smith (1992) and put it unaltered in your essay, you should add a note to the caption saying 'from Smith (1992)'. If you have made changes to the original version then you would write 'after Smith (1992)'. If you have produced a completely new version of the figure based on the data in the original, then you could write 'based on Smith (1992)'. You *must* credit the source, or you will be guilty of plagiarism.

If you use photographs that you have taken yourself, you should crop them to a suitable size so that they don't include irrelevant material or empty space around the edge. You should also consider whether sticking the actual print on to your essay page will make a clumsy mess and make the paper physically more difficult for the reader to handle. If so, consider scanning the photograph into your word-processing package, or making a colour photocopy of the whole page with the photograph on it, so that the photo becomes one with the page rather than a stuck-on object that could fall off later.

If you use a figure from some other source, you will almost certainly need to adapt it to make it fit perfectly in your essay. One thing that tutors hate is when a diagram is dropped into an essay complete with details that only apply to the original source and not to the essay. Even something as straightforward as an ordnance survey map extract usually needs to be adapted before use. Do you really need *all* that detail in your essay's map? If you need a map to show the locations of the volcanoes and the cities, don't put in a map that also shows the road network, the river system, the state boundaries and the railway lines. All that superfluous detail will obscure your point and lose you marks. If you take figures from other sources, edit them to suit your purpose; don't just regurgitate them undigested and with all the feathers still on!

And don't forget to include a scale.

## Checklist

When you have a figure ready to go into your essay, check that you can answer 'yes' to all of these questions. If you can't, then you should make changes to the figure, or to the essay, until you can.

- Does the figure have a number?
- Is it numbered correctly in sequence with the other figures?
- Is the figure referred to by its number in the text of the essay?
- Does the figure have a clear and detailed caption?

- Is the source of the figure or the data acknowledged?
- Is the figure well drawn and clearly reproduced?
- Is the figure printed at an appropriate size?
- Are key features on the figure labelled?
- Does it have all necessary scale and orientation labels?
- Is all lettering clearly printed and legible?
- Has all irrelevant or superfluous information been removed?
- Would the essay be noticeably worse if the figure were missed out?
- Is the figure adequately discussed in the text?

## Conclusion to Chapter 13

- Figures can make an important contribution to your essay. Good figures will make the essay better and earn you extra marks. Poor figures will make the essay look shoddy, may confuse or mislead the reader, and will lose you marks.
- Figures really jump out and make a first impression on the reader. An essay with a truly horrendous figure on the front page will make a bad first impression. If your essay sparkles with high-quality illustrations, neatly integrated into the text, readers will notice this and get a positive impression even before they start reading.
- You do not have to be a great artist to produce great figures. Just remember that each figure is there to make a point. Make it clearly and integrate it carefully into the text of your essay.

# 14 How much do things like layout, neatness and English matter?

## Chapter summary

Things like layout, neatness, spelling and grammar, all of which fall under the general heading of presentation, matter a great deal. They matter directly because you may be assessed on them, and indirectly because your message will come across more clearly if it is well presented. You should word process your essays wherever possible, and make sure that the text is clearly legible. The size and style of font, and the placing of blocks of text on the page, make a big difference to how your tutor will judge your essay. Syntactic sloppiness, grammatical gaffs and poor punctuation will lose you marks and destroy the readability of your essay. Getting this right is partly a question of knowing the rules and techniques, and partly a question of just taking care over your work.

## Why can't I just worry about the geography?

Eric says: *I thought this course was supposed to be geography, not English!*

When your essay is assessed it will be marked on two things: how much geography you know and how good you are at writing essays. The marks you get for *both* of these will be affected by the quality of your presentation. First, if the essay is badly presented the marker will not think highly of your essay skills. Second, if the essay is badly presented the reader is more likely to misunderstand things that you have written, to miss some of the important points that you have made, or to lose the thread of your argument and not realize what a good answer you were hiding behind the shoddy exterior of your presentation. Good presentation gets credit in its own right and also makes it possible for your geography to shine through clearly and get the credit it deserves. Bad presentation gets marks knocked off and obscures the quality of the geography that you have written.

## So what things do I have to worry about?

A lot of different things come together to make a well-presented essay. We mentioned the importance of producing a *well-written* essay in Chapter 8, but presentation involves a lot more than correct writing. Some of the most important things to get right include:

- following the specific instructions for your assignment;
- size, spacing and layout of text on the page;
- choice of handwriting or word processing;
- legibility of text;
- spelling;
- syntactic sloppiness, grammatical gaffs and poor punctuation;
- size and quality of paper;
- correct binding for submission;
- correct number of words.

## Should I handwrite or word-process my essay?

Like many aspects of presentation, the decision about whether you should word-process your essays may already have been made by the course team that set the assignment. Check your instructions carefully and ask your tutor if you are not sure. If there was no specific instruction, it is nearly always best to word-process your essay.

**Word-processing makes it easier for you to write your essay, and easier for the reader to read it.**

Word-processing helps at the writing stage mainly because of the ease with which it enables you to make both minor corrections and major structural alterations even after you have written the bulk of the essay. For example, when giving your essay a final read-through before submission, you realize it would be better to reverse the order of two paragraphs. If the essay has been handwritten, it will be a major job to make that change. It will involve either some very messy crossing out or a lengthy re-copying session. By contrast, if the essay has been word-processed, it will take just a few seconds to make the change and there will be no visible scar from your operation.

Word-processing generally makes life easier for the reader because the text is likely to be more legible than most people's handwriting. One of the reasons tutors often dread exam marking is because of the horrors of trying to penetrate students' handwriting. The only real advantage of handwriting your coursework essays is to check whether your handwriting is good enough to get you through an exam. Bad handwriting is one of the main causes of students not having the quality or detail of their exam answers fully appreciated by a tutor who has not the patience to stumble word by scribbled word through the morass of a messy essay. How would you like it if this book were handwritten?

## How should I set out the text on the page?

The main things to think about when arranging text on your page are:

- comfort of reading;
- visibility of structure;
- space all around for marker's comments.

You don't want the reader to struggle with densely packed microscopic curly text. You do want the reader to be able to see at a glance the structural breaks in your text. You do want to leave sufficient room for the marker to write in the side margins, at the top and bottom of the page, and in between lines of text. You want the person who picks up your essay to be able to read it, judge it fairly, and write comments on it.

The main things that you can control to make sure you achieve this include:

- font and font size;
- line spacing and margins;
- paragraph break markers.

We'll deal with these in the next few sections.

## Which font should I use?

The font is the lettering that your word processor uses. It is equivalent to what used to be called the typeface. Your computer will probably offer you a huge range of different fonts ranging from curly scripts that look like handwriting to more straightforward text of the sort that this book uses.

**You should use a simple clear font such as Times New Roman or Arial.**

You should avoid fancy fonts partly because they can be difficult to read, partly because they can give an unprofessional look to your work, and partly because not all word processors or printers will be able to recognize them. It would be a problem if you submitted your essay electronically and your tutor's machine could not read it.

Even using a single font there are various effects that you can apply to emphasize or highlight sections of text. For a start, you can use the font in different sizes. The most common use of this effect is to put the title of the essay in a larger font than the rest of the text. Within the text you should use a font size that is easy to read. Typically 12-point is about right. Some people prefer 10-point, but you need to allow for the fact that your tutor's old eyes might not be as fresh as once they were; 14-point is unnecessarily large and 8-point is definitely too small. Your course might have specific regulations about required font sizes, so do check your instructions.

In addition to choosing the size of font there are several other effects that you could apply. For example you can make your text **bold**, or *italic*, or ***both***. You could underline text or put it in all UPPER CASE letters. You could even use coloured text if you plan to produce the essay on a colour printer. There are many possible ***COMBINATIONS***. As a general rule you should use these effects very, very sparingly, if at all. You might use italics *very* occasionally to emphasize a word, but whenever possible it is better to structure your text in such a way that the emphasis is obvious, rather than using technical effects to achieve it. Keep your text as plain as possible unless there is a good reason to vary it.

One good reason to use an alternative font is to distinguish between different types of content in your essay. For example, you might put all of your figure captions in italics to distinguish them clearly from the rest of the text. In reference lists it is sometimes helpful to distinguish between the title of an article and the title of the publication in which it appears by consistently using italics for one item throughout the list. Another possible use would be to identify direct quotations, though this is usually unnecessary as the quotation will be identified by quotation marks. It's common to put foreign words (such as *et al.*, *per se* and *Calluna vulgaris*) in italics.

**Don't use fancy effects unless they actually help to make the essay clearer.**

## How should I space the text on the page?

You need to worry about line spacing, margins all around the text, and breaks between paragraphs. There may be rules about each of these in the instructions for your assignment, so check.

Usually you should produce your essay with 'double' line spacing. Line spacing can be 'single' (which means that there is no real gap between each line), 'double' which means that there is space big enough to squeeze a whole extra line between the lines of text, or 'one-and-a-half', which would leave a gap of about half a line's thickness between lines of text. You should produce your essay at one-and-a-half or double spacing. Try them both and see which looks best.

**Single spacing is usually too close for an essay because the eye can find it hard to track between lines and the tutor can find it impossible to squeeze comments legibly into the script.**

You should leave wide margins all around your text. This tends to make the text easier to read, and also makes space for the tutor to write comments. If you are using A4 paper, then you should leave at least 3 cm of clear space all around the text. Try re-formatting your text to experiment with different settings for the margins. Don't go so far that your text looks like a postage stamp in a snowy field. Think very carefully before expanding a margin beyond about 5 cm. You don't have to keep the margins symmetrical: you might leave more space at the top than at the bottom, and more space at the left than the right, for example.

Once you start looking at the various tricks that your word processor can do, you might start wondering about putting your text into double column format or setting out your essay as a fold-out leaflet. Alternative approaches such as these can be effective in the appropriate place, but usually an essay is not that place. Save effects like that for other types of assignment. In your essays keep it simple and keep it clear.

## How should I mark paragraph breaks?

Paragraphs are the building blocks of your essay's structure, so it is very important that they are clearly identifiable. Standard practice in writing is to break the line at the end of a paragraph and either indent the first line of the next paragraph, or leave an additional line space at the end of each paragraph, or do both. Unless you have been given a specific instruction, it's up to you to do what makes the text seem most clear. We recommend that you do both. If you only leave a line and do not indent, then paragraph breaks that coincide with page boundaries become invisible. If you indent but do not leave a line space, then paragraph breaks can be hard to distinguish from other indented text such as bullet lists or quotations.

There are other options for marking new paragraphs, such as the 'dropped capital' that begins this paragraph. Once again however, we would urge you to consider simplicity to be the better part of clarity. You will lose more brownie points from a tutor who doesn't like your 'effect' than you will gain from one who does. What you should NOT do (note our use of capitals) is leave the paragraph break completely unmarked. If you neither leave a gap nor indent the first line, then when a paragraph ends near to the end of a line the paragraph break will become invisible and your two paragraphs will appear to run together as one. This will destroy the structure that you set up in the plan, and will lead to confusion or misunderstanding on the part of the reader.

## Downright bad writing

We considered in Chapter 8 the importance of your essay being 'well written'. It is, in fact, so important that we will consider it again!

If you were to submit to a professional journal or to a book editor a manuscript that contained worthwhile material but was marred by spelling mistakes, grammatical errors and poor expression, you would have it returned with a negative response. Part of your training in geography is to prepare you for the big bad world of real life, so if you submit an essay marred by spelling mistakes, grammatical errors and poor expression, you will have it returned with a negative response. Eric wondered at the start of this chapter whether we thought this was an English course instead of a geography course. What Eric needs to appreciate is that geography involves communicating your discoveries or ideas to an audience. In order to do that, you have to speak their language. The language of geography for those of you reading this book is English. You need to use it. If you use it incorrectly your readers will (a) think that you are ignorant and (b) have a hard time understanding what you mean.

Downright bad writing is an important issue under the heading of presentation, but since we covered it earlier in the book and since we don't want to set you a bad example by being repetitive, we suggest that you take a look back at Chapter 8 if you need a reminder about how to make sure that your essay is well written.

## What if I'm dyslexic?

The answer to this question depends on your institution. Most institutions have very clear procedures for ensuring that dyslexic students, and students suffering from any other sort of disability, are treated fairly. However, fair treatment does depend on students keeping the institution fully informed about their needs. If you are dyslexic or have any other condition or circumstance that might affect your ability to present your work, consult your tutor for advice right away.

## Checklist: is your essay well presented?

- Is your essay correctly labelled to identify the course, title, date and author?
- Is your essay free of spelling errors? (Do a spell check.)
- Is your essay free of grammatical errors? (Do a grammar check.)
- Are the paragraphs clearly separated?
- Are there wide margins all around the text?
- Is the line spacing wide enough? (At least 1.5 space.)

- Is the font large enough? (12-point should do.)
- Is the font clear enough?
- Is the use of fonts consistent throughout the essay?
- Is the print sufficiently dark and crisp?
- Are the pages numbered?
- Is the paper of the right size and thickness?
- Is the paper free of creases and dog-ears?
- Is the paper free of coffee-cup stains and other filth?
- Is the paper free of the stink of cigarette smoke?

If you can answer 'yes' to all those questions your essay is probably looking OK. However, there is always scope for improvement with little personal touches. You could have a cover sheet in different coloured paper. You could include your name and the essay title in a running header on each page.

**You could decorate each page with little pink bunny rabbits and spray perfume on to the paper.**

Be careful. Don't overdo things. Remember that clarity is your primary goal. There is a time and a place for little bunny rabbits, and your essay is not it.

Finally, your institution is likely to have rules about submission of coursework. There may be anonymity numbers so that tutors don't know whose work it is they are marking. Anonymity numbers won't be much use if you have your name on every page of the essay. You may be required to include a departmental cover sheet or a submission form that is a required part of submitted coursework. Sometimes the cover sheet is also the mark sheet that your tutor will fill in. If you don't include the cover sheet the whole assessment procedure will get held up and you will be singularly unpopular. If there is a cover sheet, or any other paperwork that you are supposed to include with your essay, make sure you include it.

## Conclusion to Chapter 14

- So you've written an essay, done all the diagrams and the reference list, and you've checked that everything is nicely presented.
- You might think the next step is to hand it in.
- Not quite. Read Chapter 15, and think about how to perfect your essay before you hand it in.

# How can I perfect my essay before I hand it in?

15

## Chapter summary

When the essay is finished, go through it right away and check for errors, then let it go cold for a while and come back to it with a fresh eye. Read it aloud to somebody and see if it sounds any good. Mark it, and give it a grade using the marking criteria that we discussed in Chapter 4. Identify the things that bring the mark down from 100% and fix them. Show it to other people and see if they have any good advice. Finally, when everything is perfect, hand the essay in – making sure you follow all your department's rules about labelling, packaging and submission.

## Why can't I just hand it in now?

A weak student will hand an essay in as soon as the conclusion is written and the final item is added to the reference list. This is partly because a weak student will finish the essay very close to the deadline, and partly because a weak student will think that when the last section is written the essay is complete. A good student will reach the stage that we have reached in this book, having written the complete essay with all its sections of text, references and illustrations, and will then pause before handing it in. This is possible because the good student will have completed all the stages up to this one well in advance of the deadline. It is necessary because the essay can still be improved.

There are three stages to writing an essay:

- research and planning;
- drafting and writing;
- checking and improving.

Everybody has to do the writing stage, and it's pretty much impossible to write an essay without at least *some* preparation, but an awful lot of students miss out the final stage of checking and improving the essay after it has been written.

There are several steps that you should take after your less ambitious colleagues would have packed up and handed the thing in:

- checking the essay for mistakes while it is still hot;
- letting the essay go cold;
- reading the essay as a stranger;
- marking your own essay;
- making changes to improve the grade.

## Checking the essay while it is still 'hot'

As soon as you think the essay is finished, at the stage when you could, if you had to, just hand it in, you should immediately run a series of basic checks to make sure that everything is OK. This is the bit that people are talking about when they say, 'Always read through your essay when you've finished it.'

**Even a simple read-through usually turns up a number of silly little mistakes that would, if they went uncorrected, irritate or confuse the reader.**

Typical errors that can be picked up at this stage include:

- spelling mistakes;
- words missing out of sentences;
- fragments of text that were not deleted with the rest of their cut section;
- words typed twice twice when you paused to think in mid-sentence;
- sentences that just make no sense;
- numbers mistyped or without units ('the population of the UK is 58');
- paragraphs that accidentally run on to the next without a break;
- figures that don't have captions;
- text references that don't feature as items in the reference list and vice versa;
- and many more.

Silly little errors like these occur for all sorts of reasons. If you are word-processing it is always possible to hit the wrong key and produce an error. It is also very common for people to edit out part of the text and then forget to reword the surrounding sections to accommodate the change. Often, especially when rushing in an exam, the brain works faster than the fingers and we miss out one or two words of writing because our fingers unconsciously skip ahead to keep up with the brain. Whatever their causes, silly little errors are very common in essays that are fresh out of the oven. You need to check through, find them and correct them. Many mark-changing errors would have been easy to spot and correct if the student had just read through the essay before handing it in.

When you read through your own work, or through somebody else's work if it is clearly written, your mind tends to anticipate what is coming a few words ahead so if there is word missing out of a sentence you tend not to see it. The missing 'a' before 'word' in that last sentence could be a tricky one for me to pick up when I check through this chapter. To make sure that you actually *see* the silly little errors you should read your essay *out loud*, one sentence at a time, so that missing words or impossible sentences force you to stop reading. In an exam, of course, you will have to do this very quietly!

## Should I use the computer's spell checker?

If you are word-processing your essay, there may be a tool to help you within your word-processing package: a spell checker or a grammar checker, for example. These can be very useful as a quick way of picking up errors, but they usually don't do a very complete or reliable job.

**Foe exam pull, is ewe rote thin sentence, the spell chequer wood knot picket up because they are all correctly spelt words.**

You also need to make sure the spell checker is set to UK English, or else it may default to US settings and will 'correct' all your spellings into their American counterparts. To test your spelling and grammar software it is an interesting game to feed your essay into the programme and allow all the changes that are suggested to be made automatically. There is a good chance that your essay will have more, and more serious, errors after the checkers have done their evil work than before. Box 15.1 shows a paragraph as it was intended, as it was originally

### Box 15.1 A question as it was intended, as it was originally typed, and as the computer on which this book was typed tried to have us correct it

*As intended:*

Are Tors on Karst of interest to speleological experts?

*As mistyped with just two errors:*

Are Tors on Karst of <u>intest</u> to speleological <u>ecsperts</u>?

*Following the suggestions of the spell checking software:*

Are Tours on Carts of intent to stereological escapers?

*or:*

Are Torsos on Karts of infest to stereological espartos?

typed, and as the computer on which this book was typed tried to have us correct it. By all means allow your computer to highlight what it thinks may be errors, but don't trust it to spot them all, don't assume that something is wrong just because your checker doesn't like it, and don't automatically trust the solutions that it may offer.

## Letting the essay go cold

After you have run your initial series of 'hot' checks you should let the essay go cold. In other words, put it away and don't look at it for a while. When you are caught up in the middle of a piece of work it is difficult to judge it objectively, but if you take a break, stand back, and look at it as if it were a stranger to you, you can judge more clearly whether it is actually any good. The length of time for which you can let the essay cool off depends on how far ahead of the deadline you managed to get to the checking stage. In an exam you might be able to let one essay go cold for a few minutes while you write or check another. In a coursework essay you should aim to let it go cold for at least a couple of days. Ideally you should leave the essay for as long as possible. A couple of weeks is usually achievable if you are well organized. In order to make this possible you need to make sure you get everything done well in advance of the submission date.

Eric says: *Plan for this cooling-off period when you set up your timetable for writing the essay.*

When the essay has cooled off, take it out of the fridge, and before it begins to warm up and become familiar again start reading it out loud from the top. Ideally have somebody with you and read it out loud to them. This is the big test. If the essay is good it will be easy to read out loud even from cold: you will find it easy to read out and your listener will find it interesting and easy to understand. This is as close as you can get to seeing how your essay will appear to the person who picks it up off the pile, looks at it for the first time and gives it a mark.

## Mark your own essay

After you have read through the essay, give it a mark. This is not as easy as it sounds, so don't rush it.

**Marking essays is a complicated business.**

Look back at Chapter 4, where we discussed what the examiners would be looking for when they marked your essay. In that chapter we looked at the formal mark sheets and markers' guides that tutors often use to assess essays. If they can do it, so can you. If you can get hold of a copy of the markers' guide that your tutor will use, and one of the mark sheets that your tutor will have to fill in, go and get one now. If you can't, then use one of ours from Chapter 4. Boxes 4.2 and 4.3 were examples of the criteria that define different grades and Box 4.1 was an example of a simple mark sheet.

**Match your essay to one of the sets of criteria in Box 4.2. What grade does it correspond to?**

Does your answer show:

- detailed knowledge of the subject?
- deep understanding of all the issues raised by the assignment?
- original insights based on independent thought?
- rigorous and logical argument?
- evidence drawn from a wide range of high-level source material?
- impeccable organisation and presentation?

Does your essay contain:

- errors of style, language or presentation?
- factual errors or misconceptions?
- indications of plagiarism?

Write a 100-word assessment of your essay. Fill in one of the mark sheets. Work out what grade it would get. Make a list of all the things about your essay that pulled the mark down from 100%.

## What should I do if my essay is looking good?

If your essay gets an A+, First Class, Excellent, then you are struggling to find anything wrong with it. Of course, each of you will have your own threshold for what you think is 'good'. Some of you will only be happy with a First Class mark. Others of you will be happy as long as you are scraping in with a pass. That's your problem. If you are happy with the mark that your essay achieved on your self-assessment, then you are almost ready to hand it in. However, do be aware that your own assessment might not correspond with that of your tutor. If you have been honest and rigorous in your assessment it should be close, but are you sure you covered all the things that your tutor will be looking for? Were you as rigorous as your tutor will be? It might not hurt to have another look at Chapter 4. You might also just want to show your essay to a friend, or even to your tutor if that is allowed. After that, you can skip ahead to the section below on what to do just before you hand the thing in.

## What should I do if my essay is *not* looking so good?

It is likely that the mark you gave your essay was not as good as you would like to achieve. If you didn't give yourself full marks, that just means that there were things you thought were not perfect in your essay. Well, if you can see what the problems are you are halfway towards fixing them. Now fix them!

> The point of doing the whole cooling off and marking exercise was to help you to identify things that you could improve.

If you can make a list of things that pulled down your essay mark, then you have a list of jobs that you can do to improve the essay before you hand it in. Work your way through your list fixing one thing at a time. When you've done that, go back up this chapter a couple of sections and mark your essay again. How does it do this time?

## What should I do if it's all gone horribly wrong?

Don't worry, most things can be fixed. Go and read Chapter 21.

## Who else can I show the essay to?

Generally, the more feedback you can get on your essay before you hand it in, the better. Show it to friends. Show it to your family. Show it to complete strangers that you meet on the street. You need to know whether your essay makes sense, and other people are going to be able to judge that more easily than you. To them your essay hasn't just gone cold, it's completely out of the blue.

Your tutor will no doubt have an opinion about whether students should bring finished essays for unofficial reassurance before handing them in. Usually, tutors will be too busy to cope with a whole class of students wanting their essays pre-marked, and they would not thank us for sending you to their doors. However, if you have specific worries about some aspect of your essay, or if you want advice about whether to present something in one particular style or another, then your tutor will probably be able to help. It certainly won't hurt to ask.

## OK, it's ready to hand in. Now what?

When the essay is really finished and you are ready to hand it in, you need to make sure that it is submitted in accordance with the rules and that you haven't made any silly last-minute mistakes. Here are some of the things to check.

- Have you got the right submission date?
- Are you supposed to put your name on the front of the essay?

- Do you have to put the course title, module number or tutor on the front?
- Should you put the date on?
- Are there specific rules about having a title page, or a cover?
- Are the pages supposed to be stapled together?
- Should you number the pages?
- Where are you supposed to hand it in?
- Are you supposed to get a receipt for the essay?
- Are you required to submit two copies?
- Are you required to keep an extra copy for yourself?

The answers to all of these questions, along with any other rules, regulations and petty annoyances that your department puts in your way, should be explained in your course handbook. Make sure that you know what you are supposed to do and that you do it. After all the work you've put into the essay you don't want to have it disqualified for incorrect submission.

If in doubt, just make sure that the essay is clearly labelled so that people will know what it is if it gets mixed up in the wrong pile. Make sure that it is firmly fixed together so that pages don't go missing. If you're worried, put a running header or footer on each page with your name or identification number on it. Page numbers will also help if the sheets come apart and get mixed up.

A lot of students like to submit their essays in little plastic wallets. This might help to keep the paper clean and prevent dog-ears, but when a tutor gets a pile of 150 essays all slipped into little plastic wallets how much fun do you think it is taking each essay out and slipping each essay back? At 5 seconds per re-insertion, 150 essays makes about 12 minutes of stuffing essays into plastic wallets. Give us a break. At least, don't expect your essay to come back in the same wallet it went in. In fact don't even expect to see your wallet again. Most tutors generate a big pile of leftover plastic wallets each year. There are tutors who could have quit academia and retired on the proceeds of used plastic wallet sales.

## Conclusion to Chapter 15

- Don't forget to hand it in.

# 16 How should I handle exam essays?

*Chapter summary*

In an exam you don't have much time, you can't check your sources and you can't ask for advice. Thorough preparation before the event, and good time management during it, are therefore essential. Although your exam essays will be designed to achieve the same things that your coursework essays did, there are some special circumstances to take into account in an exam.

## Do I have to write essays differently in exams?

Much of the advice that we have offered so far in this book has been organized in a way intended mainly to be useful when you are writing coursework essays. Most of this advice applies equally well in exams, but there are some big differences between writing coursework essays and having to write an essay in an exam. The most important differences include:

- the amount of time that you have to plan and write the essay;
- the amount of reference material that you have available to help you;
- the amount of advice and support that you can get while you are writing;
- the amount of anxiety that you may experience during the assignment.

All of the general principles of what makes a good essay are the same for an exam essay as for a coursework essay, but some of the problems that you face in an exam are different from those that you faced in coursework, and there are some differences in the range of solutions available to you. In this chapter we'll give you some advice that applies specifically to writing essays in exams.

## How should I prepare for an exam?

We suggested earlier in the book that the key to writing a good essay is good preparation, and that most academic tasks are generally straightforward once the groundwork is done properly. This certainly applies to exams. If you are well prepared you will probably do well; if you are poorly prepared you will probably do badly.

Eric says: *'There is a strong positive correlation between thorough preparation and exam success' (Knight and Parsons, 2003). Discuss.*

There are a lot of things that you should do in the weeks leading up to the exam to help to make sure that you go into the exam well prepared and confident. The most fundamental element of good preparation is to have acquired a deep understanding of core issues and a detailed knowledge of case studies through diligent pursuit of the course. Whether or not you have that background, you can do the following.

- Make sure you know what the exam is about and what the format will be.
- Think about what the examiners will be looking for (revisit Chapter 4).
- Gather and revise material on relevant topics.
- Summarize your core understanding into a condensed form.
- Look at old exam questions.
- Think about how what you know could be applied to past questions.
- Write some practice answers and have them informally assessed.
- Become confident that you understand the topic.
- Become confident that you know lots of relevant information.
- Get plenty of sleep, exercise and fresh air; eat well and relax.
- Plan how you will manage your time in the exam.

Preparation for exams is a little bit outside the scope of this book, so let's skip forwards to what you do once you get into the exam. If you want to read more about revision techniques and exam preparation, there are plenty of books on those topics that you could consult. Look under 'study aids' in the library or bookshop.

## How should I divide up my time in the exam?

One of the reasons most commonly given for exam failure is: 'I ran out of time.' Well, if you run out of time you deserve to do badly.

**Time management is one of the skills you are being examined on.**

A student who can write three essays in 2 hours (or whatever the exam requires) deserves to do better than one who can't.

Time management depends on knowing how much time you have, what jobs you need to do, and how long each job will take. If you have read the previous chapters in this book you will have a good idea of all the things that are involved in writing an essay, but when we have discussed them previously we have allocated days or weeks to each stage. You can't do that in an exam.

The major jobs are:

- choosing which questions to do;
- thinking about the topic;
- planning each question;
- writing each essay;
- checking each essay 'hot';
- letting each essay go cold and making improvements.

At the start of an exam you should make yourself a timetable and make a note of the specific times by which you have to finish each stage. In fact, if you are a good student you will have worked out this timetable in advance, as we suggested in the list of bullet points in the previous section.

## The simplest approach to time management

The simplest approach to time management would be to divide the total exam time by the number (*n*) of questions that you have to answer, and to spend one *n*th of the total time on each question. In a 3-hour exam with three questions you would make sure you moved on to the second question after 1 hour, and so on. Box 16.1 shows what this simple timetable would look like, but there are problems with it.

### Box 16.1 A very simple, but not very sensible, exam timetable

2 pm Start exam, start first question

3 pm Start second question

4 pm Start third question

5 pm End of exam.

This timetable doesn't allow time for choosing which questions you will do. If your exam is one where you get to choose which questions to do, you will need to take some time off the total for choosing questions. Since this is a very important part of the exam (see Chapter 5) you need to allocate quite a lot of time to it. You could allocate as much as 10% of the total exam time. In a 3-hour exam, where you have to choose three titles from a list of nine, that would mean about 15–20 minutes choosing which questions you should do. That only works out at about 5 minutes per question, but if you take all that time out of your first question's hour, it skews your timetable badly. Take the time out of the timetable before you allocate time per question (Box 16.2).

**Box 16.2 A simple exam timetable allowing time for choosing questions**

2.00 Start exam, start choosing questions (15 minutes)

2.15 Start first question (55 minutes)

3.10 Start second question (55 minutes)

4.05 Start third question (55 minutes)

5.00 End of exam.

Success with this approach depends simply on making sure that you complete all the stages of writing the essay within the allotted time and making sure that you move on to the next question when the time is up. In other words, stick to your timetable. It sounds simple, but an awful lot of students mess it up.

What most often goes wrong is that students think they know a lot more about one question than another, so they tweak their timetable to give more time to a favourite question. Don't do that. Always allocate time evenly between questions.

Remember:

- The marks you earn per minute of writing in each essay diminish with time.
- Extra minutes spent at the end of an essay earn marks very slowly.
- Minutes spent at the start of a new essay earn marks very quickly.
- Difficult essays, where you don't know much, take a lot of time to write.
- If you don't finish your last question you are more likely to fail the exam.

Eric says: *You* must *stick to your timetable.*

## The sophisticated approach to time management

Rather than doing all the jobs for question 1, then moving on and doing all the jobs for question 2, a lot of people prefer to do the plans for all the questions together, then do the writing for all the questions, then come back and do the checking for all the questions. One advantage of this approach is that you are less likely to get carried away with one question before thinking about the others, so you are more likely to write all the essays that you are supposed to. If you are doing things this way, you can't use a simple timetable like the ones in Boxes 16.1 and 16.2. You need to think about how long to allow for each of the different parts of writing each essay. How long on planning? How long on writing? How long on checking? Box 16.3 suggests a sensible breakdown of times for a 1-hour essay.

## Box 16.3 Sensible breakdown of times for a 1-hour essay

| | |
|---|---|
| *Preparation* (20 minutes): | |
| Choosing the question from a list | 5 minutes |
| Thinking what the answer is | 5 minutes |
| Drawing up a plan | 10 minutes |
| *Writing* (30 minutes): | |
| Introduction | 5 minutes |
| Middle bit | 15 minutes |
| Penultimate section | 5 minutes |
| Conclusion | 5 minutes |
| *Checking* (10 minutes): | |
| Read-through | 5 minutes |
| Time for corrections/additions | 5 minutes |
| *Total* | 60 minutes |

If we then spread these times through a whole 3-hour exam, we would end up with a timetable like the one in Box 16.4.

From this timetable a few things might jump out at you:

- You can sit in the exam for an hour before actually starting on your first essay.
- Only half of the time is spent actually writing the essays.
- You have half an hour of checking and improving time at the end of the exam.

Things aren't as rushed as people make them seem. Relax.

Eric says: *Exams are not races. The 3 hours are to see how well you can answer all the questions, not to see how fast you can fill the pages.*

**Box 16.4 Sensible breakdown of times for a 3-hour three-essay exam**

| | | |
|---|---|---|
| *Preparation* (60 minutes): | | |
| Choosing the questions from a list | | 15 minutes |
| Thinking what the answers are | | 15 minutes |
| Drawing up three plans | | 10 minutes |
| *Writing* (90 minutes): | | |
| First essay | | 30 minutes |
| Second essay | | 30 minutes |
| Third essay | | 30 minutes |
| *Checking* (30 minutes): | | |
| Read-through | | 15 minutes |
| Time for corrections/additions | | 15 minutes |
| | *Total* | 180 minutes |

## What should I do at the start of the exam?

Find your seat, check that you have been given the right bits of paper, and make sure that all the pens, bottles of water and hay fever tablets you might need are within easy reach. Make sure you can see the correct time. Take a moment to relax. Don't rush. Think about all the preparation you've done. Remember all our advice about choosing questions, making detailed plans and leaving time to read through your answers. When you are allowed to do so, take a look at the exam paper. Don't let it see any fear in your eyes. Only take a quick look, then look away again, relax, and let what you have seen sink in.

Is it the right paper? Have the topics that you were hoping for come up? Don't worry if it looks impossible; you know that when you take a few minutes over it you will find questions that you can answer well.

At this stage you will almost certainly see that most of the other students have picked up their pens and are writing ferociously. If you could see their paper you would probably see that most of them have started writing their first essay while you are still in the 'relaxing before thinking about it' stage. This is normal. They probably don't know what they are doing and they're writing rubbish. It is interesting watching students taking exams.

**The students whom you know to be complete idiots with no knowledge or understanding of the subject are usually the ones who start writing first.**

Poor souls. They should have read our book. Now, don't worry about them any more. Breathe normally, relax your shoulders and start working through your jobs.

Make a quick timetable. Choose your questions. Think about the answers. Construct plans. Check the time against your timetable. Stay relaxed. Take breaks and look into the distance occasionally. Remember everything you've learned. Be in control.

The 'choosing questions' stage is crucially important. Make sure you follow the instructions at the top of the exam paper. Don't pick two from section A if you are supposed to answer only one from each section. Look at Chapter 5 again, where we went through this in some detail. If you go right through the question paper and can't find enough questions that you can do, just relax for a moment, look away, cool down and take another look. Often, the initial panic and stress of the situation makes people think they can't do questions that are in fact quite simple once they've calmed down. It is important to pick all your questions right at the beginning of the exam rather than waiting till after you've done the first two before choosing the third. If you pick them all at the start you can have the later ones stewing away in your subconscious while you write the early ones. It's the exam equivalent of memorizing a coursework title and then doing nothing (on that title) for a while (Chapter 6).

## What do I have to do differently in exams from coursework essays?

Exam essays and coursework essays are trying to do pretty much the same things, so you aren't really aiming for anything different in an exam from what you were aiming for in coursework: a well informed, clearly presented answer to the question, fulfilling any instructions given in the title. The only real differences are differences of style and presentation.

- You don't normally need a reference list.
- Your essay will be handwritten rather than word-processed.
- You may need to compress your message into a shorter space.

We discussed references in exams in Chapter 12, so you might want to take another look at that now.

We stressed the importance of legible presentation in Chapter 14. This applies particularly to exams where you are handwriting: you may be in a hurry (although you shouldn't be if you are well organized) and you might have to make alterations by means of crossings-out rather than by word-processor cut-and-paste. If the examiner can't read what you say, you won't get marks for it.

Eric says: *I knew nothing, but I wrote so scruffily that the examiners wouldn't have been able to tell. I still failed the exam.*

You may need to compress your essay into a shorter space in an exam than in coursework because you may not be able physically to write as long an essay in the 30 minutes or so of writing time that you have in an exam as you could in the unlimited time available in coursework. The solution to this is to be direct and concise. Don't say less, just say it less long-windedly. You can make more use of techniques like bullet points to cut down on writing time. Sacrifice fancy wording. However, do not sacrifice clarity, do not sacrifice legibility, and do not sacrifice material. Do not sacrifice planning time. Do not sacrifice checking-through time. The best way to make sure you squeeze the best essay into the available time is to plan carefully. You can even use your plan as a figure in the essay: 'Figure 1 is a flow diagram indicating the main interrelationships relevant to this question.'

# How do I handle 'seen' questions?

Some tutors like to set seen questions. Not all tutors agree that seen questions are a useful contribution to exam assessments. They might say that, to some extent, a seen essay is just a coursework essay that you have to write out again without study aids in the exam. This reduces the exam element to the level of a memory test. If all students could remember exactly what they prepared, the best 'coursework' would produce the best answer. If students remember differently, the best memory will win. Despite their obvious flaws, you may nevertheless encounter them. One approach to seen questions, therefore, is as follows.

1. Prepare the essay in the usual way as if it were coursework.
2. Memorize it.
3. Regurgitate it in the exam.

**The problem with regurgitating prepared answers comes if you can't memorize things perfectly.**

Most people can't memorize things perfectly, so we need another approach, and this is the approach your examiner will expect you to take. Basically, remember the most fundamental points and the key examples. Think of a seen question as an opportunity to focus your revision. You now know what material to read, and you can do all your preparation before the exam. You go in confident that the topic you have revised will come up. If you have done your preparation well you should be able to answer a seen question really well. That means you will feel confident that you have a good answer in the bag when you come on to any unseen questions on the paper. But remember to give a seen question no more than its fair share of the exam time. If it's worth a third of the marks don't get carried away with all the material you have prepared and spend two-thirds of the exam time on it. A good idea is to write and remember a really good introduction that will also serve as a plan as you write the middle bit of the essay. If nothing else, just try to memorize the key points, essentially the plan. Regurgitate the plan in the exam, then re-write the essay from scratch using your plan as a starting point. Even if you have no memory, if you have the plan you should be able to build an essay from scratch as if it were an unseen question. However, not everybody will be starting from the same position of ignorance.

## What should I do at the end of the exam?

- Do not leave early.
- Check through each essay 'cold' at the end of the exam.
- (You should have checked each essay 'hot' as soon as you finished it.)
- Make improvements to each essay as appropriate.
- Check that you really have answered the right number of questions.
- Make sure you have filled in the front of the answer paper properly.
- Follow the instructions for submitting your paper to the invigilator.
- Do not leave early.

Eric says: *Do not leave the exam early.*

## Can I leave the exam early if I've finished?

If things have gone well, you will probably find that you are busy right up to the end anyway, but if you do finish early it means that you have not been kept busy for as long as the people who set the exam expected. This is probably your fault, not theirs. If your tutor was sitting with you in the exam when you thought you were ready to leave, do you think your tutor would be able to suggest some improvement to the essays? Some additional example? An extra case study? An alternative point of view on your second answer? Surely there must be something. Try doing a self-assessment and give each of your essays a mark. Now think how you could improve that mark. Take another look at Chapter 15.

**If things have gone badly, leaving early will not help; staying longer might do.**

If you feel you've just had enough and can't go on, just take a break. Put your pen down. Lean back in your chair. Close your eyes for a few minutes. Let your mind wander. Think about next weekend. Write a letter to a friend. If you need to, ask permission to go to the toilet. Just don't walk away from the exam! After 10 minutes you might feel better and be able to do something to turn things around. If you've left the exam you won't get that chance. If you fear all is lost and hope is gone, try to remember the advice from Chapter 21. This will be easier if you have read Chapter 21 before the exam, of course. Preparation is everything!

Eric says: *'Prepare in haste, repent at leisure.' What do people mean when they say that?*

## Checklist: 10 exam commandments

1. Thou shalt not answer the questions that thou wouldst have been asked, but only those that thou perceivest before thee.
2. Thou shalt not render violence unto the grammar and spelling of the English language, nor forsake the virtues of legibility; for the examiners shall not find guiltless he or she that doth so.
3. Thou shalt not forsake the instructions at the top of the page, nor depart from the allowance of time that thou hast allotted to each question.
4. Thou shalt not steal out of the appointed place before thou hast checked through all thine answers and corrected that which is not right.
5. Thou shalt not beat thy breast shouldst thou be lacking in time: for it is better to outline thy main ideas and important details in note form than to allow thy page to remain even as the driven snow.
6. Thou shalt not covet thy neighbour's speed of pen: for it is with thy self that thou art struggling, and not with thy neighbour. And indeed it may truly come to pass that he hath been labouring in vain over mere dross and vanity and codswallop.
7. Thou shalt not attempt to crush the spirit of thy neighbour, nor dash your neighbour's hopes by saying after thou hast been examined, 'Verily what a piece of cake it hath been.'
8. Thou shalt not make unto thee upon thy shirt cuff, nor upon thy palm, any likeness of that which hath been in thy textbook. Neither shouldst thou consult the fruits of thy neighbour's toils, for the examinations and assessments committee shall convey thine iniquities unto the disciplinary bodies, which shall shew no mercy unto them that keepeth not these commandments.
9. Thou shouldst on the appointed day cultivate a clear mind. It is not meet that on the day preceding thou learnest new things, but rather revise thy knowledge and gather together whatsoever thou needest for the trial that is before thee. Neither on the night preceding shouldst thou consume alcohol or other strong elixirs, nor indulge in riotous living, but rather have thyself a glass of milk and an early night.

10. Heed the advice of thy counsellors and tutors, for they do labour hard and strive to make good thy passage from this place, for which passage we all shall offer hearty thanks and praise.

## Conclusion to Chapter 16

- Your exam will go OK as long as you know your stuff, do exactly what the instructions say, and remember to write good essays.
- This book offers some advice on how to write good essays, but if you don't know your stuff, reading this book will probably not be enough to help you. Preparation is everything.
- When you're in the exam don't rush, don't panic, don't leave early. Methodically apply everything you have learned about writing essays, be careful and check everything.
- Remember that the exam is your opportunity to show yourself to your best advantage. Don't mess it up.

17

# How do I prepare a poster presentation?

### *Chapter summary*

**Whereas an essay is designed to be read page by page on little sheets of paper, a poster is designed to be put up on the wall and seen all at once. Posters should not rely on large amounts of text to get their points across, and should be based much more around visual material. They need to be extremely succinct and well organized to get their message across in a very short time, and can use different techniques from those that you have previously used in essays. Creating a good poster is a very specialized skill. You can use our checklist to see whether your poster is a good one.**

## What are posters?

A poster is a means of communication. Like an essay, it presents information and ideas in the form of a response to a specific question or instruction. However, whereas an essay is designed to be read page by page on little sheets of paper, a poster is designed to be put up on the wall and seen all at once as a single object, often with the person who made it standing in attendance. Whereas an essay is a bit like a journal paper that is likely to be read carefully by somebody who has settled down to study it, a poster is likely to be looked at in a crowded room by people moving from poster to poster talking about the things they see.

Posters therefore need to:

- be very clear, and readable from arm's length;
- have a strong visual appeal;
- make an immediate impact on the viewer;
- be easily digestible, concise and direct;
- provide maximum information in minimum viewing time.

Poster presentations are an important part of most academic conferences. They are used as a way of allowing large numbers of people to present information about their research without occupying long stretches of the conference timetable. All the posters are put up in one big room, and conference delegates can wander around and look at them either in a timetabled 'poster session' or during the lunch breaks and intervals between spoken presentations. The person who made the poster will usually be standing beside it to take questions and discuss the work. Often, people produce posters about 'work in progress' where it is not yet appropriate to complete a final paper. A poster is therefore by nature ephemeral compared with its more permanent cousin, the published paper.

Characteristically, therefore, posters are:

- state-of-the-art;
- controversial;
- hot-off-the-press and bang up-to-date.

Compared with an academic paper in a journal, a poster is intended to be much more easily digested, and is intended for reading in a specific environment. Your essays may aim to be like journal papers, but your posters have to serve a different function.

## Why do I have to produce posters in my course?

You are asked to produce posters for several reasons. First, you might need to produce posters in a variety of situations after you finish your studies. Posters are used not only in academic conferences but also in a wide variety of commercial and professional situations. Second, posters are a very good way of forcing you to exercise some very specific skills that are valuable in many forms of communication. These include both general communication skills such as brevity and clarity, and technical skills such as manipulating text and graphics on a computer. Third, the use of different techniques in your assessment allows you to demonstrate skills in a range of areas, and to ensure that weaknesses in any one area (such as essay writing) do not overwhelm your whole degree result. Finally, doing posters just adds a bit of variety, both for you and for the people who mark your work.

## How will my poster be assessed?

Posters are typically assessed in one of two different ways. Make sure you find out which of these will apply to your poster. One method is for the tutor to take in the posters in the same way as essays, and to take them away to mark just like any other assessed work. The other method is to arrange a 'poster session', just like one that might occur at a conference, where posters are put on display, authors are invited to stand beside their offerings, and people roam around looking at the work that has been presented. These poster sessions can be thrown open to the whole department, so that your work is exposed to a wider audience: this helps to create the atmosphere of a real conference. Whichever of these approaches is used in your course, the same criteria will be used to assess your work. We need to consider what those criteria are.

## What will the examiner be looking for in my poster?

Just like an essay, a poster allows you to demonstrate your expertise in the subject as well as your technical ability in this particular form of presentation. Some of the things that the examiner will be looking for in judging your poster are just the same as those that examiners would look for in an essay (take a look back at Chapter 4).

Core questions that markers ask about any assignment include the following.

- Do you understand the title and appreciate the key issues?
- Do you have a good knowledge and understanding of the core material?
- Have you worked independently beyond course material?
- Can you organize a complex range of material from different sources?
- Can you present logical arguments with sound evidence?
- Can you produce well presented work?
- Can you work to a deadline and follow instructions?

Posters present particular opportunities and pose particular problems. They are designed for a specific purpose, so the assessment will be based on your ability to recognize and achieve that purpose.

Assessment criteria specific to poster presentations include the following.

- Is the poster visually appealing?
- Does the poster make points with brevity and clarity?
- Can the poster be read easily from a distance when it is on display?

In the same way that there are usually markers' guidelines and formal mark sheets for essays, there are often specific guidelines and marking sheets for posters.

**You should look at a copy of the assessment sheet that your institution uses so that you know what your tutors will be looking for.**

Your tutor should explain this to you anyway. If you have not had the specific expectations of your poster assignment explained to you by your tutor or in a course handbook, ask! In the meantime, we can offer you an example. Box 17.1 is an example of the kind of mark sheet that a tutor might use to assemble feedback on your work.

### Box 17.1 The sort of mark sheet that a tutor might use to assemble feedback on your poster

**POSTER ASSESSMENT SHEET**

Student ..................................................................

Title of poster .......................................................

| **Marking criteria:** | **Mark:** | **Comments:** |
|---|---|---|
| Relevance of content to title | | |
| Quality and accuracy of content | | |
| Range of sources used | | |
| Visual design and layout | | |
| Legibility and clarity | | |

**General comments:**

**Overall mark** ........ % Grade: ........

Marked by: ....................................

Date: ....................................

Box 17.1 covers the sort of points that tutors will consider if they take the poster away to their offices to mark. If they do the marking in a poster session, they might add other criteria to their list, such as how the student handled questions from viewers at the poster session. If you are having a poster session, make sure that you know whether your performance will be assessed, or whether your mark depends on the poster alone. If you do have to cope with a poster session, the section on poster sessions later in this chapter will be especially relevant to you.

## So, what should I do differently from my essays?

Because posters aim to do a different job from essays, and because the assessment will reflect the extent to which your poster succeeds in this, you need to do some things differently in your poster from what you would have done in an essay.

First of all, your poster needs to be visually clear. When we discussed essays we suggested that visible blocking of the text could help the readers to find their way around. In posters, that advice moves right up to the top of the list.

**The viewer should be able to see the structure of your poster at a single glance.**

It must be instantly obvious which bits of the poster are the start, middle and end of your message. It must be instantly obvious which figure goes with which heading. It must be obvious at first glance whether you are presenting a list, or a pair of conflicting viewpoints, or a description of a feature. These things should be apparent even without reading the text: they must be obvious from the organization and layout of the material. In an essay you created signposts by using particular phrases at the start or end of a paragraph. In a poster you can make visual signposts. You can quite literally use arrows to point the way, or you can label each section of text with a number, so the reader knows in what sequence to read the sections. Take a look at the section on how to lay out your material a little further on in this chapter.

Second, you need to be much more brief and direct in a poster than in an essay. Focus only on the most important points. There is no space for the kind of discussion that is appropriate in an essay.

**In a poster you need to stick to the essentials.**

You should replace lengthy sections of text with bullet-point lists. If a graph, photo, map or diagram can be used to replace words, use it. If you can cut out text by using subtitles and an annotated map, go ahead. The rules of 'good style' that applied to an essay are different here. There are still rules, of course: don't think you can get away with bad English. In fact, bad English will be more apparent when there are fewer words to hide it amongst.

Third, posters need to be based much more on illustrative material and much less on text than essays are. Partly this is so that you can achieve your aim of being brief and direct, and partly it is so that you can produce a poster that is a visually attractive object when it is on display. Try to tell your whole story in pictures. Whereas an essay is text supported by illustrations, think of a poster as illustrations held together with a bit of text. You can even shift most of your text into the figure captions.

Eric says: *This is what we like: less writing!*

## How should I arrange the materials on my poster?

Because academic staff in your department will have produced posters for conferences and they won't want just to throw them away afterwards, there's a good chance there will be posters on display in your department. It's bad form to keep taking the same ageing poster to lots of different research conferences, so people tend to give their posters dignified retirement on the walls of their offices. There may also be posters put up for visit days for prospective students, details about the Socrates scheme that your department is involved in, etc. Look at these posters to get ideas. But look at them critically. Don't assume that just because they are on the walls of your department they meet all the criteria for a good poster. You can probably do better than many of them.

Like an essay, there are some basic components that we would normally expect to see in a poster:

- title;
- list of authors;
- short summary;
- series of key points or stages in your story;
- conclusion;
- acknowledgements and references.

The positioning of these items on the page is largely a question of individual taste in design and layout, but you should be careful. Do not adopt a scheme that is outrageous unless you are sure it will work. Maybe for your first poster you should practise a tried and tested method, then develop your innovative approaches once you have demonstrated competence in the basics.

We'll assume we are dealing with a standard rectangular shape (say A2 or A3 size) in landscape orientation (longways across). In a straightforward approach the title of the poster should go as a prominent headline in large text across the top. This should be big enough to be read from a distance so that people will be drawn to your poster from elsewhere in the room. The list of authors should normally go immediately beneath the title in slightly smaller text. The short summary could then go in a wide shallow box immediately underneath the list of authors. You might see an immediate problem. If the text is short then all the boxes of text could end up very wide and shallow and those who view your poster will have to swivel their heads through about 180 degrees to get from one bit to the next. In general people will expect to read from top left to bottom right, so you can make that assumption in the layout of your poster. Or you can have columns, but these are rather formal divisions of the poster and are probably better reserved for situations where you need them.

The core material for the poster then goes in the big empty space in the middle of your page. It should be organized into convenient bite-sized chunks rather than spewed out as a single spillage. Each chunk should have its own heading, and could be numbered to indicate the appropriate order for reading. If you have two major points to make (for example, the opposing points of view in a two-way debate) then split the central space into two sections and put the two sets of material side by side in two big boxes. If you have three or four sections, try using columns for the material. If your material is in the form of a description of a complex system, fill the centre of the poster with a flow diagram that illustrates it.

**Remember that the design of your layout should reflect the structure of your content.**

These central sections must be well illustrated to the extent that visual material should dominate the text. People should literally be able to *see* what you are talking about without having to put on their reading glasses and read text (a minimum of 24-point text is a good rule of thumb). You can put the detailed back-up and explanation for each figure in its caption, and summarize the significance of the material in lists of bullet points. We said you should not use headings in essays because they were big ugly signposts. You should use them in posters. A poster *is* a signpost!

As with an essay, you need to take the viewer to a destination, so you need a conclusion or a terminus at the end of your poster. If you have cited sources, as you should have done, you will need a reference list. The conclusion and the references need to be clearly labelled and readily recognizable, and the logical place for them is underneath the central section, at the bottom of the poster. The references are the only section of the poster where you could get away with using smaller text than can be read at arm's length. If people are really interested in chasing up your sources they can lean in a little to read the small print. Still, you shouldn't go below 12-point as an absolute minimum, and 14-point is usually safer, even for the references.

The scheme we have described will work for a poster where the structure runs from the top of the page to the bottom. There is nothing to prevent you from designing a structure that runs from left to right, or one that spirals in towards the centre. You will get the best marks if you find an appropriate match between the content of your poster and the style in which you present it.

**It's up to you how you design your poster, but remember the golden rules: keep it clear and don't confuse the reader.**

When you start work on your poster, especially if you are preparing it as part of a group, spend some time 'brainstorming' and sketch out lots of layout ideas on paper in rough. Then compare each idea with the 'do and don't' poster checklist at the end of this chapter.

Eric says: *Can I just write it as an essay and then paste the pages on to a big board?*

The worst way to prepare a poster is to write it as an essay and then just paste it up on a big board. Posters and essays are different beasts. If you don't understand that, re-read the last few pages of this book until you do!

## How should I handle a 'poster session'?

If your tutor takes the idea of poster assessment to its logical conclusion, you should be assessed in a 'poster session'. This can take the form of a mini-conference, where all the posters from the group are put on display, tea and sandwiches are provided, and everybody wanders around looking at each other's posters. During the course of this the tutor assesses each poster. The tutor may also look at the posters again in private afterwards, of course, if only to fill in the paperwork without a bunch of students watching.

**Part of the assessment in a poster session might be based on what you have to say for yourself when people ask you questions about your poster.**

This is sometimes used as a first introduction to public speaking for students who have not yet presented formal talks. It's a nice soft introduction.

The main things you need in order to do well in a poster session are:

- a good poster;
- a good understanding of what you've said in your poster;
- additional material as back-up for when people ask you questions;
- a willingness actually to speak to people when they ask questions;
- an ability to listen when people give opinions.

If you have a really poor poster, you might be able to bluff your way through to some extent with good people skills, but this is not the recommended course of action. Prepare for the poster session by making sure you have a good poster. If you have additional materials that you couldn't squeeze on to the poster, take them with you. When somebody asks you a question you can whip out the extra material to help to elaborate your answer. It will also show your tutor that the poster is only the tip of the iceberg of your knowledge and understanding.

Eric says: *Let me remember that one: 'This poster is only the tip of the iceberg of my knowledge and understanding.'*

## Checklist: is my poster OK?

If you have you done a poster and you want to check whether it is up to scratch, use this checklist.

- *Make sure you're following the regulations:*
  - Does the poster conform to the assignment instructions?
  - Does the poster conform to specific regulations about size and shape?
- *Make sure you can answer 'yes' to all the following:*
  - Is all the material on the poster directly relevant to the title?
  - Can all text, including figure captions, be read from 1m away?
  - Can all the figures be read from 1m away?
  - Have lengthy sections of text been replaced by lists, tables and summaries?
  - Is the poster dominated by maps, graphs, photos and other illustrations?
  - Is it immediately obvious in what order readers should look at things?
  - Is there a short, prominent summary of the basic point?
  - Is the main message clear from just the title, figure captions and conclusion?
  - Is your poster pleasing to the eye?
  - Do you have an introduction, a conclusion, and a reference list?
- *Make sure you can answer 'no' to all the following:*
  - Does any block of text in the poster take more than 30 seconds to read?
  - Would you have to read more than 25% of the text to understand the poster?
  - Does it take more than 2 minutes to understand the key point?
  - Is the poster dominated by text?
  - Are items on the poster squeezed too close together?

## Conclusion to Chapter 17

- Posters are pretty straightforward if you follow the simple rules of clear communication. Don't just write an essay and pin it on the wall!
- Be careful not to copy posters that you see prepared by academic researchers without comparing them with our checklist. Most academic researchers are not actually very good at preparing posters. Many of them just write the poster as if it were a paper for a journal and then paste the material on to a big board! Whereas journal papers are usually refereed to make sure that the worst examples never get published, the same is not true of posters. Some of the posters displayed around the walls of your department may be examples of how *not* to prepare a poster.
- Why don't you put your tutors to the test? Ask whether you can do a mock assessment of one of their posters as a tutorial exercise! If you've got this far through this book, you should be able to put them on the spot.

18

# How do I prepare and deliver a verbal presentation?

*Chapter summary*

The ability to stand up in front of a group and speak sensibly is a very valuable skill. Talks require different skills from essays, but they are still about knowing your stuff and communicating clearly. It is hard work to keep an audience interested and to get your point across effectively in a talk, but with careful preparation you can achieve excellent results even if you are nervous.

## What is a verbal presentation?

A verbal presentation is basically a talk, but talks come in a variety of guises. Giving directions to the pub is a verbal presentation, and so is giving a formal lecture to 500 people. Geographers should possess both of these skills. Verbal presentation can also be important in your exams if you are faced with a viva voce exam, and most courses involve verbal presentations from students in their coursework.

**Verbal presentation features in the Geography Benchmark Statement issued by the Quality Assurance Agency, and forms a part of most students' experience in geography.**

Verbal presentation is less common than it used to be in geography coursework. Rising student numbers and diminishing staff resources in many departments have made it increasingly difficult to make space in the timetable to accommodate student talks. This is a shame, because the ability to stand up in front of a group and speak sensibly is a very valuable skill.

## What are the main differences between written and spoken presentations?

In a written presentation, whether it is a poster or a web page or an essay, readers can move through the presentation at their own pace. They can go back to re-read bits if they wish and they can skip ahead through the boring bits whenever they choose. In a spoken presentation the speaker dictates the pace. People in the audience don't get a second chance if they miss something, and they can't skip ahead through the boring bits.

From your point of view as the presenter it is essential that:

- you keep the audience interested every single second;
- you make points so clearly that nobody wishes they could 'read it again';
- you give just the right amount of time to each item you show or discuss.

Eric says: *Your readers will decide the pace at which they read your essays. You decide the pace at which they hear your talks!*

# What talks might I have to do?

At one extreme you might have to stand up by yourself in front of a huge group of other students and give a lengthy formal lecture of some kind. For most students this is the closest thing to hell that their course can offer. It needn't be. Your tutors are hardly superheroes, but *they* manage to stand in front of a crowd and speak. There must be a simple trick to it that you can learn.

At the other end of the scale a verbal presentation might just involve explaining your work to somebody in front of your poster in a poster session (Chapter 17), contributing to small-group discussion in a tutorial, or giving a short talk to a few students in a seminar.

**Unless you work yourself into a state about them in advance, these types of presentation are no more arduous than chatting with friends or ordering beers at the bar.**

If your department handles things kindly you will gradually work your way through the range of different types of talk as your course progresses. You will get used to speaking informally in a small group in your first semester tutorials, you may give short formal talks to that familiar group in your second year, and by your final year you may be giving more substantial presentations to larger groups. If you are lucky you will finish off your studies with a chance to exercise your verbal skills in the viva.

# Why do they make me do talks?

They make you do talks because it's good for you. It's a worthwhile skill to learn: 95% of human beings don't like to get up in front of a group and speak in public, and 95% of people have to do it from time to time either as part of their job, or at a wedding, or at some public forum. Those people who have had some training and experience are less likely to mess it up than those who haven't. Since fear of messing up is the biggest cause of anxiety when you have to speak, knowing that you are a trained professional is a big advantage. If ever you get a job in teaching or lecturing, or in management or local government or most of the other professions that geographers occupy, being able to stand up and speak to a group will be an advantage.

They make you do talks also because it's a very good way for you to develop, and for your tutors to assess, certain aspects of your ability. To do a good talk you need be able to:

- communicate clearly;
- organize material concisely;
- use evidence effectively to support well balanced arguments;
- remain calm under pressure.

If you have read the preceding chapters of this book certain items on that list will be beginning to look very familiar. This is what your degree is all about: communication, organization and clarity. Giving talks takes these standard items and packages them in a different format with different technicalities of presentation. Giving talks is a good test of the skills that you should be developing throughout your coursework. Why do geographers need to learn to remain calm under pressure? Well, there's the exam coming up, after that there's job-hunting, a career, life . . . Remaining calm under pressure is an important graduate skill.

## What will the examiner be looking for in my talk?

Examiners are always looking for the same stuff:

- knowledge and understanding;
- insight;
- good organisation;
- clear communication.

These were the same things we talked about when we discussed posters and essays, and we'll mention them again when we cover web pages and group work. What we need to think about is how the examiner will try to recognize these things in your talk. What evidence will give the clues as to whether you have these skills?

The examiner will be looking at two sets of things:

- academic content;
- presentation skills.

Academic content really is the same stuff that we have discussed previously. Box 18.1 lists some of the key things that the examiners will be looking for. It may remind you of Chapter 4.

## Box 18.1 Some of the academic qualities that the examiner will be looking for in your poster

- Detailed knowledge of the subject.
- Deep understanding of all the issues raised by the assignment.
- Original insights based on independent thought.
- Rigorous and logical argument based on evidence.
- Evidence drawn from a wide range of high-level source material.
- No factual errors or misconceptions.

Presentation skills are the specific techniques by which you ensure that your talk does its job of communicating effectively. We will discuss this in more detail in the section on how to present your talk later in this chapter, but in the meantime take a look at Box 18.2, which is an example of a mark sheet used for assessing verbal presentations. The list of subheadings under *Presentation Style* gives you an idea of what the examiner will be looking for.

## Box 18.2 An example of a mark sheet used for assessing verbal presentations

*Student's Name:* *Topic of Presentation:*

*Academic Content:*

Breadth of reading

Understanding of reading

Critical engagement with reading

*Presentation Style:*

Pace

Quantity/length

Structure/clarity

Interest

*Best features of the presentation:*

*Suggestions for improvement:*

Overall Grade: Signature of Tutor: Date:

Box 18.3 is an example of a criteria-based markers' guide specifically designed for verbal presentations. It identifies the specific criteria, both academic and technical, that determine what grade your presentation will achieve. This gives you a detailed insight into what the examiners will be looking for. There was another example of this sort of guide in Box 4.3. Take a look back at that one, too.

## Box 18.3 An example of a qualitative marking guide for verbal presentations

| Class | Description of presentation |
|---|---|
| I | **Academic Content:** Must show detailed knowledge of the subject, deep understanding of all the issues raised by the assignment, original insights based on independent thought, and rigorous and logical argument based on sensible interpretation of evidence and examples. Examples must be drawn from a wide range of high-level source material. Must not contain factual errors or misconceptions. Demonstrates substantial effort. |
| | **Presentation:** Well structured. Clearly spoken with a wide range of vocal intonation. Runs to time without rushing or being too slow. Engages the audience's enthusiasm. Extremely polished and confident. Audio-visual aids are relevant and of excellent quality. The mechanics of the presentation do not distract the presenter or the audience during the talk. Demonstrates excellent preparation. |
| 2.i | **Academic Content:** Must show knowledge of the subject, understanding of the main issues raised by the assignment, insights based on independent thought or reading, and rigorous and logical argument based on sensible interpretation of evidence and examples drawn from a range of source material. Must not contain serious factual errors or misconceptions. Demonstrates good effort. |
| | **Presentation:** Well structured. Clearly spoken with some range of vocal intonation. Runs to time without rushing or being too slow. Engages the audience's interest. Polished and confident. Audio-visual aids are relevant and of good quality. The mechanics of the presentation do not seriously distract the presenter or the audience during the talk. Demonstrates careful preparation. |
| 2.ii | **Academic Content:** A relevant presentation that recognizes the key issues raised by the assignment but may reveal shortcomings in knowledge or |

| | |
|---|---|
| | understanding, and may include errors, omissions and some irrelevant material. Must demonstrate basic knowledge and understanding of course material, and must use evidence to support arguments, but may lack evidence of independent thought and critical analysis. Shows some signs of effort. |
| | **Presentation:** The material is presented coherently, but there may be minor weaknesses in structure or clarity, and the speaker may need to rush or dawdle in order to run to time. The speaker may not engage the audience's interest throughout the talk, and audio-visual aids may be of inferior quality. The performance gets the main points across, but is a little rough around the edges. Shows some signs of preparation. |
| 3 | **Academic Content:** A poor presentation that reveals limited knowledge and understanding of the subject but nevertheless recognizes the main issues raised by the assignment. Must provide a relevant answer, but may include a few serious errors, omissions or irrelevant material. May lack evidence of independent thought or reading beyond basic course material. Arguments may be superficial and lack evidence. Limited effort evident. |
| | **Presentation:** The presentation hinders the communication of material. There may be significant weaknesses in structure or clarity, and the speaker may not run to time. The speaker may use a dull monotone and read directly from a prepared script. The speaker may lose the audience's interest, and audio-visual aids may be misleading or inadequate. The presentation may appear to be under-rehearsed. Limited preparation evident. |
| Fail | **Academic Content:** Little evidence of understanding. Factual material is limited and/or incorrect, and structure barely discernible. May suffer from faulty reasoning and/or miss the point of the presentation altogether. Insufficient effort has gone into this. |
| | **Presentation:** Embarrassing for audience and presenter alike. Insufficient preparation has gone into this. |
| | **Plagiarism:** This grade will be applied to work that shows evidence of plagiarism. |

From the markers' guide and mark sheet in Boxes 18.2 and 18.3 you can draw out a number of key things that the examiners will be looking for in terms of presentation:

- clarity;
- timekeeping;
- polish and confidence;
- good audio-visual aids;
- mechanics of presentation that do not distract from content of talk.

Achieving these things will come partly from the structure and content of your talk and partly from the mechanics of your presentation. We will discuss each in turn.

## How do I make my talk well organised?

The organization of a talk is not much different from the organization of an essay. You need to say what you are talking about (start with the title); explain the key points that you are going to cover (the introduction); run through the core of the material point by point (the middle bit); and then arrive at your conclusion. In a talk you also need to leave time for questions and discussion at the end. The order of play is thus:

1. Introduction
2. Middle bit
3. Conclusion
4. Questions.

The key to putting together a good talk is good planning. You need to do all the same background for a talk as you do for an essay.

- Get to grips with the question.
- Get to grips with the answer.
- Research your material.
- Organize your structure.
- Prepare a detailed plan.

In doing all of this you need to be constantly aware of how much time you have been allocated for your talk, and therefore how much material you will be able to squeeze in.

**There is no point gathering enough material for a 10-part epic on the fall of Troy if you only have 5 minutes in which to speak on it.**

In preparing your plan, think about all the issues we raised when discussing essays and posters. All the same elements of good clear structure with signposting and logical fluency apply to a talk. You still have to answer the question directly. You still have to stick to the point. You still have to use evidence to back up your assertions. However, what you should not do is just write an essay and then read it out. There would be several problems with that approach. First, having a complete 'script' to read out word by word usually leads to a very poor presentation. We'll discuss that in the next section. Second, if you write an essay you are likely to produce lengthy sections of detailed discussion that might not hold the interest of an audience and might take too long to read out anyway. What you should do instead is transform your detailed plan into a set of 'speaker's notes' that you can use as prompts when you speak. We'll talk about them later in this chapter.

With a talk, rather like a poster, you need to keep your audience interested by:

- starting with a good clear first statement;
- focusing your material on the key points;
- keeping it snappy and not lingering too long on any one point;
- making the signposting absolutely clear;
- finishing with a good clear closing statement.

## How should I perform my talk to make it 'well presented'?

Actors sometimes say that the key to acting is just to remember your lines and not annoy the audience. The same is pretty much true of giving a talk. The trick is knowing how not to annoy the audience.

Generally your audience will get annoyed if:

- they can't hear you;
- you bore them;
- you have annoying habits like saying 'erm' a lot;
- they don't understand you;
- they lose the thread of your argument;
- they can't read the slides you put up;
- you don't make eye contact while you talk;

- you go on too long;
- you don't go on long enough;
- you don't talk about the right stuff;
- you trail off weakly at the end without a good closing.

We could make this list longer, but it should already have made the point that there are many ways to annoy an audience. You need to avoid them all. A good general pointer is that you should always remember that the members of the audience are giving up some of their valuable time to sit politely and listen to your talk. It's only polite that you should prepare carefully and give them the best talk that you can. If you don't, you will be thought very rude.

## What are the 'top tips' for presenting a talk?

### 1. Speak very clearly

Speak loudly enough to be sure that everybody can hear you. Speak slowly. Enunciate each syllable carefully. Pause for a second between each sentence. It might sound odd to you as you are doing it, but it will sound normal to the audience as they try to hear every word you say. Remember, they need time to hear, process and think about each word.

### 2. Speak naturally and don't read from a script

If you read out from a fully written script you are likely to speak in a dull monotone with unnatural intonation and emphasis. It won't sound normal and it will be very dull. Speak naturally, making each point 'in your own words', using your written notes just to remind you what each point is before you deliver it.

> Don't ever just read your talk straight from a script.

### 3. Use body language and eye contact

If you just stand at the front and speak, your performance will be very dull and people will find it difficult to follow what you say. In natural speech, body language makes a huge contribution to successful communication. You should use gestures, expressions and movement to assist your words. Be animated. Eye contact is vital: look around the audience as you speak. You must seem to be interested in them. Don't just look at your notes, or out of the window, or at your computer screen. Feel as if you are having a real conversation. Talk *to* them, not *at* them.

> Don't ever just read your talk straight from a script.

### 4. Use visual emphasis

When you make a point, be sure that people realize you are making a point. People find it much easier to follow your speech if they can see it as well as hear it. Visual emphasis can be a bit of body language, holding up an object for the audience to see, or even a simple 'x' on the blackboard drawn with emphasis as you say, for example, 'We have *one* main target'. The 'x' will focus people on your point, add variety to your presentation, and help to make your presentation seem natural and fluent. 'That . . .' (tap the board with your finger) ' . . . is visual emphasis.'

> Don't ever just read your talk straight from a script.

### 5. Use visual aids

People find it much easier to understand what you mean if you can *show* them as well as *tell* them. Visual aids can be anything from a projected slide or a map on the wall to a hardware model or a handout. However, visual aids are only any good if they work. Make sure your slides are clear. Make sure the lettering on your overheads is large enough to be read from the back of the room. Make sure your visuals are visible for long enough. Your audience will be unhappy if you whip away a chart before they have had a chance to digest it.

Eric says: *If a visual aid is not worth showing for at least 20 seconds, it is not worth showing at all.*

Visual aids are also useful to the speaker. If you put a list of bullet points on screen for all to see, you can work through them without having to look at your own notes. A visual aid for the audience is a secret prompt for the speaker. Visual aids can also be used for drawing people's attention away from yourself if you are a nervous speaker. Plunging the room into darkness and directing audience attention to a screen is a great confidence booster!

### 6. Appear confident

One of the worst things for an audience is to be faced with a speaker who appears really nervous. It actually distracts people from listening to what you say. It's quite OK (and perfectly normal) to be nervous; you just need to learn not to show it. We'll discuss this later in the chapter.

### 7. Don't fidget

Fidgeting distracts the audience and gets in the way of your intentional body language. Sometimes we have bad habits we don't even know about. You will see lecturers who constantly pace, or fiddle with their hair, or say 'OK?' at the end of every sentence. Normally they don't even know they're doing it. Watch out for those habits and control them. OK? An interesting exercise is to video yourself giving a talk. Watching the tape can be a painful but educational experience.

### 8. Keep signposting

Remember, your audience doesn't have your text in front of them so they don't know how far through you are. You need to keep them informed about your progress with little phrases such as 'Moving on to the second of my three main points' or 'My final point before we summarize and reach a conclusion is . . .'. You can reinforce this by having the programme of the talk on display or giving a handout.

## What should my 'speaker's notes' look like?

What you *don't* want is a complete script where every sentence is written out in full. First, it will probably lead you to a dull reading; and second, you will probably lose your place, not be able to work out where you were, and the whole talk will go wrong.

Eric says: *Don't just read your talk straight from a script.*

You need prompts for each point to remind you what the point is, but once reminded you should then tell the audience about the point 'in your own words'. This will lead to a much more natural and fluent presentation. Box 18.4 gives you an example of how a 'full script' can be reduced to speaker's notes. Notice how you can re-tell the story from the same notes in different versions. They all make the same points in the same order, but they all sound just a little bit different. That's because the speaker just makes the words up fresh each time the talk is presented. The notes keep things broadly on target and can be used to remember specific dates and numbers, but the actual words spoken are 'made up' fresh on the spot.

Using notes like this has several benefits, as well as keeping your talk on target. The fact that you have to do a tiny bit of mental organizing to translate from the notes into spoken sentences (as opposed to just reading the sentences straight off a script) makes you go just a tiny bit slower than you would if you were reading. This is a good thing, as most speakers tend to go too fast. The fact that you are making the sentences up on the spot rather than reading a pre-prepared speech will ensure that you are using 'spoken' English rather than 'written' English. When you write, you tend to use certain styles that just don't seem right when you speak them. For example, you would want to *say* 'they're' and 'it's' whereas you would normally *write* 'they are' and 'it is'. That is why essays sometimes seem a bit stilted when they are read out loud. Try reading that last sentence out loud and you will see what we mean.

You will have noticed in Box 18.4 that our notes were printed extra large. It is important that you should be able to see your notes very easily, and you should not have to pore over them to work out what they say.

## Box 18.4 An example of how a 'full script' can be reduced to speaker's notes

**Notes for short introduction to talk:**

Hello.

Today – urban redevelopment.

3 points:

1. General problem of redevelopment (UK).
2. Stoke-on-Trent (case study: post-1970).
3. Compare: Stoke (success), other cities (problems).

**Spoken version:**

Good morning, everybody. In today's talk about urban redevelopment I am going to cover three main points. First I will discuss the general problem of redevelopment in the UK. Next I will look at the specific case of Stoke-on-Trent, where a great deal of redevelopment has taken place since 1970. Finally I will examine how the success of Stoke's redevelopment contrasts with problems that have occurred with redevelopment schemes in other cities.

**Alternative wording:**

Hello, everybody, thanks for coming to my talk today. I'm going to talk about urban redevelopment. There are three main points that I want to cover. First I will talk about the general problem of redevelopment as it has affected the UK. After that I would like to go through a case study with you: we'll look at redevelopment in Stoke after 1970. At the end I will take some time to compare Stoke, which has a successful history of redevelopment, with the less positive stories of some other cities in the UK.

**Another alternative wording:**

Make up your own. You've got the notes, you know what you want to say, just work your way through the notes saying the points they remind you of. Try it.

**Make your notes short, clear and direct, and don't squeeze too many to a page.**

Some people like to use index cards, with each short section of the talk condensed on to one card. Others like to use standard A4 paper, but if you do that you must make sure that you only put a small amount on each page. Box 18.5 gives you another example of speaker's notes, this time with additional prompts for when to use visual aids. You could even add prompts for yourself to use particular bits of body language, or to pause for a moment. You could just write a reminder at the foot of each page of notes: 'Speak slowly.'

Box 18.5 Speaker's notes with additional prompts

**Have first slide showing at start**

**Eye contact**

Hello

Today – urban redevelopment.

**Show overhead 1 (3 points list)**

3 points:

1. General problem of redevelopment (UK).
2. Stoke-on-Trent (case study: post-1970).
3. Compare: Stoke (success), other cities (problems).

**Show second slide**

**Pause while people look**

## Can I base my talk on computer projection?

One way of providing prompts for yourself and visual aids for the audience at the same time is by using PowerPoint or a similar on-screen presentation. We say more about this type of presentation in Chapter 19. Presentations of this type are really just a fancy way of presenting visual aids, and if used appropriately they can be very helpful. However, we should issue a few words of warning if you plan to follow this route. You may have already experienced on-screen presentations from your lecturers, so you may already have been on the receiving end of some of the pitfalls.

Remember:

- Don't cram too much information into each slide.
- Don't allow the slides to replace the talk.
- Don't give your whole talk in one medium. Use other media to break it up.
- Don't rely on technology to the extent that a power cut would shut you
- down.
- Don't read your slides to the audience.

# How do I combat nerves?

There are three excellent ways of combating nerves when you have to give a talk:

1. Be well prepared and rehearsed so you know it will go well.
2. Act calm and you will become calm.
3. Set up distractions so you don't feel under the spotlight.

Good preparation, as we have mentioned once or twice earlier in this book, is the key to success. Preparation for your talk involves:

- getting to grips with the topic;
- planning the talk;
- writing the notes;
- preparing all the visual aids;
- practising your talk.

Practising the talk is especially important. Practice will make sure that you are familiar with the material, and that you are used to making up sentences from your notes. It helps if you memorize just the first sentence of your talk. This will give you something to focus on in the moments before you have to start speaking, and by the time you have recited your memorized introductory line you will have broken the ice and those first-minute nerves will have evaporated.

As part of your rehearsals you should as a bare minimum do the following.

- Give the talk to yourself, your dog or your teddy bear out loud with all the
- visual aids.
- Check that the timing is right – not too long or too short.
- Check that you are audible from the back of the room.
- Check that your visual aids are visible from the back of the room.

If you have the time, technology and friends available you should do the following.

- Give the talk to a group of friends.
- Video yourself giving the talk (and watch the tape!).
- Pull yourself together, eliminate the bad habits and do it again.
- Keep doing it until you feel confident.
- If possible, check the room that you will be talking in and the equipment you will be using so you know where all the switches are. If you can't find the switch to turn on the overhead projector when you need it, all of your confidence will disappear – probably never to be seen again.

However well practised you are, and however confident you may be that you are on top of the material and that the talk is well prepared, you will probably feel nervous. This might be the kind of nervousness that keeps you awake at night for weeks before the event, or it might be a little flutter of butterflies just as you stand up in front of the class. It's natural. It's something to do with evolution, apparently. It is not a sickness or a psychological disorder. It is something that you can learn to deal with.

**One of the best ways of dealing with nerves is to put on an act.**

Putting on an act does two things. First, it conceals the worst effects of your nervousness from the audience, which is a good thing. Secondly, it actually seems often to have the effect of concealing your nervousness from yourself. It sounds silly until you try it, but for most people it really does work. Act confident and you will start to feel confident. Many of the world's great actors are very shy people. Many of your lecturers, whom you see standing up talking in front of large classes, are very nervous about speaking in public. They cover it up with an act. They *act* like a lecturer. Or, in your case, you act like a student who is relaxed and confident giving a talk. Watch confident speakers and imitate them. You should put on this act at the practice stage, as well, so that your 'on stage' persona becomes familiar to you before you use it in public.

As a final precaution against nerves you can set up your visual aids and other paraphernalia in such a way as to draw the audience's attention away from you. This can be dangerous, as you do not want to distract the audience from what you are saying, but you can at least stop them all staring at you and making you feel nervous at the start. For example, you can make sure that there are plenty of visual aids for them to look at. You can give them a handout to look at. You can show a pretty picture slide just as you start the talk. In addition you can give yourself little things to do, like cleaning the board or adjusting the lights, so that you don't feel self-conscious when you want to insert a short pause to slow the pace of the talk.

## What if it all goes wrong and I forget what I'm saying or lose my nerve?

Refer to Chapter 21 for emergency assistance.

## Checklist: is my talk well prepared?

- Does the talk cover the specific questions and instructions in the assignment?
- Does the talk head towards a small number of clearly identified conclusions?

- Is the academic content of good quality?
- Have I prepared a clear and succinct set of speaker's notes?
- Do I have adequate visual aids that are clear and legible?
- Have I practised the talk out loud all the way through?
- Does the talk run to exactly the right length?
- Have I left time for questions at the end?
- Have I prepared to the point that I feel confident?
- Do I know the layout of the room and the equipment?

## Conclusion to Chapter 18

- If you know your stuff and communicate clearly your talk should go well.
- Talks are very different from essays, and require special preparation. Good preparation is the key to success.
- Some people get nervous about giving talks. If you get nervous work around it and don't let it spoil your talk. Nerves wear off with preparation and practice.

19

# How do I prepare web pages and other types of presentation?


*Chapter summary*

**There are many different types of presentation that you could be asked to prepare in addition to those that we have discussed so far. The most common include: abstracts, technical reports, newspaper articles, web pages and other computer-based presentations. These different types of presentation demand a variety of skills in the use of different media, but you always need to focus on good content and clear communication.**


## What other types of presentations might I have to do?

It is an explicit aim of most courses in geography to employ a variety of modes of assessment in coursework and exams. Lecturers can be pretty inventive people, so there is no real limit to the types of presentation that you might be asked to do. The most common types of coursework presentation that are required in addition to essays, posters and talks include:

- web pages;
- other computer-based presentations (e.g. PowerPoint);
- abstracts;
- professional reports;
- newspaper articles;
- press releases.

## Why might I have to do these different types of presentation?

There are two main reasons for employing these various modes of assessment. First, they give you a different medium in which to show your strengths. If you were at art college and all the assessment was based on still-life watercolours, you'd think it wasn't a fair way of assessing your ability as an artist. So you can think of all these different modes of assessment as the equivalent of charcoal drawings, collages, metal sculpture, etc. The second reason is to give you skills that you may find useful later in life, when you have a job. Good as essays are at assessing all the sorts of things we have talked about so far, when you get a job it's pretty unlikely that anyone will ask you to write an essay. But they may ask you to prepare a press release, or to write a technical report.

## Do the same principles apply to all these types of presentation?

The same basic principles do apply to all these different forms of presentation, because they are all exactly that: forms of presentation. When you present material, you are trying to communicate. When you are trying to communicate, you want to be clear, you want to get your point across, you want to be convincing and you want to look as if you've done your homework and know what you are talking about. If you have worked your way through this book, you should be comfortable with that by now.

The same principles apply in what the examiners will be looking for. They will want to see that you have done your homework and know what you are talking about, that you have some sensible ideas, and that you can communicate effectively and get your point across.

Eric says: *So really it all boils down to knowing your stuff and communicating clearly.*

## What are the key differences between types of presentation?

When you send a telegram, it doesn't matter if you have a speech impediment. When you make a telephone call it doesn't matter if you can't spell. Different modes of communication require different skills. You can communicate the same message with each method, but the techniques are different, and you might need to organize your material in a certain way to suit the technique.

**The different types of presentation that you do in geography will force you to think about communicating in different ways.**

You need to think about:

- the intended audience (who are you talking to?);
- the available space or time for communication (how long will they listen?);

- the technical requirements of the medium (computers don't take biro);
- the purpose of your message (what are you trying to achieve?).

In this chapter we will look at some of the most frequently used types of presentation in geography and discuss the specific requirements of each.

## What matters in a web page?

Eric says: *A web page is like a poster on steroids: an all-singing all-dancing visual feast of information.*

You will almost certainly have used web sites in your course to obtain information. So, think about what helped you get what you wanted out of each page and what irritated you. It won't surprise you to find that the things you come up with are the things we have been on about all through this book: worthwhile content and clear structure. Don't let the exciting possibilities of the medium make you forget that you have a message to communicate. Because of the way web pages work, you might think that you can scatter your material all over the place and just let the user hop around at random. This is not the best approach. For a start, if users hop around at random they might not see all of your material. They might get lost and not come back to your main points. The users are free to roam, but you need to guide them. You need to provide structured navigation and a site map so that they can find their way around. Navigation provides the structure of your presentation when you make a web page. Your visitors must not get lost.

**Navigation through your site needs to be clear, simple and easy to follow.**

You need to have a 'Home Page' or 'index' where the user begins. This serves the same function as the title and introduction in an essay: explaining the main point of the site and signposting the way that the user can explore your content. You don't want too much text: a web page should be a bit more like a poster than an essay, using plenty of visuals. From the index, users should have links to each of the key areas on your site. Each of those areas will have its own front page, from which there will be links to pages of detail at the bottom of the hierarchy.

Maintaining your site as a hierarchy – with the Home Page at the top, key sections below it, and details within each key section below – makes your site intuitively easy for users to navigate. An alternative approach is to use a linear structure where the reader is directed from page 1 to page 2 to page 3 and not allowed to hop around outside the route. This is easy for you to arrange, but rather defeats the purpose of having a web site, which is to allow flexibility of movement. If you produce a web page that has a linear narrative like an essay you won't get many marks for structure. You can experiment with different structures, but a simple hierarchy is a good place to start. Box 19.1 shows an example of a simple hierarchical structure.

## Box 19.1 An example of a very simple hierarchical structure for a web site about agriculture and industry in China

*Level 1:*

*Page 1* – Introductory page (index) with links to pages 2 and 3.

*Level 2:*

*Page 2* – Agriculture, with links to pages 4, 5 and 6.

*Page 3* – Industry, with links to pages 7, 8 and 9.

*Level 3:*

*Pages 4, 5 and 6* – with details about aspects of agriculture.

*Pages 7, 8 and 9* – with details about aspects of industry.

Each individual page on your site must also be clear and easy to follow, with a structure of its own. Each page needs a title, content and navigation links to other levels in your hierarchy. Most good sites use a common layout for each page, so that the whole site has a certain 'look'. For example, you should stick with the same set of fonts, use similar backgrounds and arrange material in similar patterns on each page. You can also have a prominent 'navigation bar' that features on every page and contains links to the home page and to the key sections of the site. If the navigation bar incorporates a logo or graphical version of the site's title, it does a good job of unifying the appearance of all your pages. Some web-authoring software packages will help you with this by allowing you to define a 'template' that will keep all your pages looking the same.

There are lots of things that can go wrong in web pages and annoy the user. You need to avoid these. Major culprits include large image files that slow down loading times, and add-ons such as java scripts or other flashy gizmos that might require the user to have specific software on their computer. Remember, technical wizardry that looks good on your machine may look very different, or not work at all, on your tutor's machine.

Eric says: *Keep it as simple as possible. The less technical wizardry there is, the less there is to go wrong.*

Web pages have one huge advantage over other forms of presentation. They allow you to set up links that will zap the user from one place to another within your material, or to other sites outside your own. To make the most of your web page you need to take advantage of this opportunity. Within your own site you can exploit it through careful design of navigation. Externally you can link your site to relevant materials elsewhere on the internet.

**You can use internet links a little bit like references in an essay, providing evidence and case studies for your own points.**

However, remember that the internet offers up a lot of rubbish as well as good sites, so be careful where you link to.

Finally, your ability to do all of these things depends to some extent on your ability to manipulate the technology of web pages. In the same way that you need to be able to write English and type in order to create an essay, you need to be able to produce hypertext documents to make a web page. There are different ways of handling the technology. You might write HTML (hypertext mark-up language) code and input it directly to your computer, or you might use a WYSIWYG (what you see is what you get) software package that writes the code for you. If you have been asked to submit work in the form of a web page, we assume that you will have been given some basic technical training. Oh, and one last reminder: don't forget to load up all the gifs and jpegs as well as the html files! It's a common mistake.

Eric says: *Learn how to use your package or language, then use it to communicate clearly and effectively.*

## What matters in other types of computer-based presentation?

Computer-based presentations are becoming increasingly common even when web pages and internet connections are not involved. As technology develops, computers are simply being used to replace more traditional techniques. For example, many lecturers have now abandoned 35 mm slides and overhead transparencies and have replaced them with computer-based presentations projected on to the old slide screen at the front of the lecture theatre via an LCD (liquid crystal display) projector. Whereas lecturers used to turn up at the lecture room with a slide carousel and a pile of papers, they now turn up with a laptop and a miniature LCD projector. In fact, many lecture rooms are fitted with built-in projectors already connected to computers on the lecturer's bench.

**Computer-based presentation is becoming the norm for verbal presentations.**

It is important to decide when designing a computer-based presentation whether your computer-based material is there to support a verbal presentation or is intended to be a free-standing presentation that a user can run without your being present. If your presentation is primarily verbal, with computer-projected material incorporated as a visual aid, then the same rules apply that we discussed when we covered verbal presentations in Chapter 18. If your presentation is free-standing, then it effectively becomes rather like a web page but without the internet linkages and, usually, with a linear structure that leads the user directly from stage to stage in a predetermined order. The rules that apply here are the rules of clarity and simplicity that we applied to web pages.

## What matters in a technical or professional report?

A technical report is a document prepared to make a case about a specific problem or proposal. As geographers are often professionally involved in planning, geography students are often set the task of writing a technical report in the form of a planning document. For example, you might be asked to write a report of the type that a planning officer would prepare to present a case to a planning committee concerning the provision of traffic-calming measures in a residential area.

The important point about a professional report is that it is designed to be read and understood by people who are not experts in the field of the report. Those who have commissioned the report have asked for a recommendation that is based on an expert's expertise. The job of the report is to make such a recommendation and then to explain its basis and justification in a way that does not require the level of expertise in the subject that is needed to write the report. This is different from an essay, where you can assume that the reader knows as much about the subject as you do, if not more.

**Professional reports differ from essays in that the main aim is to provide information that will inform a practical decision.**

A planning report, for example, needs to be balanced and objective and yet it must be persuasive and move towards a recommended course of action. Whereas an essay can 'sit on the fence', a planning report has to make specific recommendations for action to be taken. The report may be prepared for a commercial client and has practical significance as well as academic interest. You should remember that the planning issues may be sensitive and controversial, and that there may be financial and social consequences of your recommendations.

In contrast to the discursive arguments typical of an essay format, technical reports are usually based upon more succinct, 'bite-sized' pieces of information. Reports normally have section headings, bullet points and numbered paragraphs, and avoid long sections of text. The reason for this is that it is not intended that the whole report will be read by everybody interested in the report. If the report is on the environmental impact of building a new runway at a major airport, there may be chapters on the impact on the local bird life and on the houses that will need to be demolished. An individual reader might be interested in only one of these issues and not want to wade through the sections on other topics to find their own area of interest.

The language in a report should be formal, avoiding slang, casual abbreviation and use of the first person. Illustrations may be appropriate, but keep them simple. Some references will be necessary, but they will probably be fewer than in a conventional essay and are more likely to be presented as footnotes rather than in the Harvard style that we recommended for essays in Chapter 12. Professional reports should not have any poor English.

Your report should have a logical structure that presents an unfolding story as the reader progresses through the document. This is achieved by going from the general to the specific, with the background material preceding the technical detail that is presented as evidence, and with conclusions and recommendations following logically from the evidence and the argument that you have presented.

A typical format for a report could include the following headings:

- executive summary – key evidence and recommendations;
- introduction and outline of main issues;
- development of factual review and main argument;
- possible courses of action and their implications;
- recommendations;
- appendices (technical data and references).

## The executive summary

This should be no more than 5% of the report, and in a long report quite a lot less. It's designed to tell the person who commissioned the report the answer to the question and the key reasons for that answer. The rest of the report provides a justification for this summary just as, in an essay, most of the text provides a justification for the content of the opening paragraph.

## The introduction

This defines the problem being investigated and the context of the investigation. It may give the limitations to the report. A surveyor's report on a house you are planning to buy may say in this section that there was no ladder available so the surveyor did not look in the loft. If the report concludes that the house seems good value for money, you need to bear in mind that that conclusion is limited by the fact that the surveyor had no opportunity to report on the dry rot in the roof timbers.

## Development of the factual review

Here the material that forms the basis for the final recommendations/conclusions is given. It includes how the data were collected, the data (though they may be given in summary tables with the full data in the appendices) and the analysis of the data.

### Box 19.2 A mark sheet designed for assessment of planning reports

PLANNING REPORT MARK SHEET

Student name/number:

Tutor/Examiner:

**Structure**

Relevant to title?

Key points identified?

Topic covered in depth?

Good conclusions and recommendations?

**Content and Argument**

Detailed?

Accurate and correct?

Supported by evidence?

Logically developed?

Persuasive?

Original and creative?

**Sources**

Appropriate?

Sufficient?

Adequate referencing?

**Style**

Fluent?

Succinct?

**Presentation**

Attractive/effective layout?

Effective use of illustrations?

**Grammar/Syntax/Spelling**

Free of errors?

**Other Comments**

**Overall Mark:**

### Possible courses of action

A report is often commissioned to enable a decision to be made among various possible courses of action. This section should consider the implications of the available options in the light of the evidence of the previous section.

### Recommendations

If the report makes specific recommendations then this final section needs to make clear why these options, rather than others that might have been available, are preferred.

Box 19.2 shows a mark sheet designed for assessment of planning reports. If you are writing one, check it against this mark sheet.

## What matters in an abstract?

An abstract is a summary of an article. It usually appears at the front of the article to enable potential readers to decide whether the article will be interesting to them, and can also be reproduced as a free-standing item, without the article attached, to tell people what the article is all about. Most geography students will have to produce an abstract when they do their geography dissertation.
We explained how to produce abstracts for dissertations in the companion volume to this book, *How to Do Your Dissertation in Geography and Related Disciplines*. If you have not looked at that book, and you have to write an abstract, we recommend it highly! Sometimes, abstracts are set as short assignments in their own right. You will be given an article or book and asked to produce an abstract of it.

The key things to aim for in an abstract are:

- brevity;
- clarity;
- completeness.

You have to be brief because there is usually a very strict and tiny word limit. Sometimes you might be limited to as few as 100 words. You have to be clear because you need to communicate a lot of material in a tiny space. You have to be complete, because that is what abstracts are all about: they take a whole article and reduce it to its essentials without missing anything out. They abstract the essence. When people have read your abstract they should know all about the article you have abstracted. You can't miss stuff out.

Abstracts are most commonly written for research articles, and they have to cover:

- the aim of the research;
- the reason the research was needed;

- the methods by which the research was carried out;
- the results that were achieved;
- the significance of the results with regard to the original aim.

The reason that an abstract covers those points is because those are the important points about a research article. If you are asked to write an abstract, just figure out what the important points are in the article and summarize each one in a single sentence.

## What matters in a newspaper article?

Newspaper articles are different from any of the types of presentation we have discussed elsewhere in this book because they are not 'academic' presentations. If an essay is written for somebody who probably has as much idea of the answer as you do, and a professional report is written for somebody who knows sufficient about the problem to know what questions to ask, a newspaper article is best written for somebody who hasn't even thought about the topic before. Your task is to make that person find the topic sufficiently interesting to want to read your article. In essays and professional reports you are guaranteed an audience. In newspaper articles you've got to find your own. The general principle that you need to communicate clearly still applies, but many of the rules of academic writing go out of the window when you write a newspaper article. For example, you would never use the phrase 'that goes out of the window' in a formal essay, but it would be OK in a newspaper article. In academic writing you have to provide well-documented evidence for everything you say. Newspapers don't have to do that. In academic writing you should develop a fair and well-balanced argument that considers alternative points of view. Newspapers don't have to do that, either.

At the top of this chapter we said that you needed to think about who you were writing for and what you were trying to achieve. With a newspaper article you are writing for a very different bunch of people, and trying to achieve some different goals, from those that you were worrying about in your academic writing.

The main goals of a newspaper article are:

- to make money;
- to keep the writer in a job;
- to help sell copies of the newspaper;
- to help sell advertising space within the newspaper;
- to provide information;
- to entertain;
- to push a particular point of view.

The basic technique by which journalists achieve these goals is to take a very particular viewpoint on an incident and report it in such a way as to generate a predetermined response. There are several ways of doing this. For example, when the *Titanic* was struck by an iceberg, it could have been reported in many different ways. Look at the alternative headlines in Box 19.3.

### Box 19.3 Alternative newspaper headlines following the sinking of the *Titanic*

- *Titanic* sunk by iceberg on maiden voyage, hundreds drowned (*Daily News*)
- Iceberg from Greenland drifts south of Newfoundland (*Greenland Gazette*)
- Transatlantic mail delayed by shipping incident (*The Post*)
- Negligence of shipping company causes unnecessary deaths (*Law Daily*)
- Famous diamond lost at sea (*Insurance Weekly*)
- Disaster team call for improvements in hull design (*Shipwright International*)
- Fewer seats available on next month's transatlantic schedules (*Travel Today*)
- Baby dies while mother looks on (*Women's Herald*)
- Environmental impact of mid-Atlantic debris (*The Green Paper*)
- Local man in boating incident (*Stoke Gazette*)

These headlines demonstrate one aspect of the golden rule of writing newspaper articles: target the story to your audience.

## What matters in a press release?

The task of a press release is to persuade people like journalists or TV editors that you have information that would interest their readers or viewers. You have to sell your story. Because it has to grab somebody's attention it has to be short and to the point (like an abstract), but it has to be written for somebody who knows nothing about the subject. Look back to the goals of a newspaper. A press release has to convince somebody that your story will do those things.

Some rules for press releases are as follows.

- Sum up the story in the first sentence, or paragraph.
- Explain the broader significance in the second sentence or paragraph.
- Write in the present tense – the news is *now*.
- Make sure your press release answers the question, 'So what?'

## What if they ask me to do some other type of presentation?

By now this should be easy:

- Make sure you know your stuff and do what the assignment requires.
- Plan carefully and organize your material into a logical structure.
- Communicate your material clearly using an appropriate style.
- If in doubt, check your handbook and ask your tutor.

Eric says: *I begin to see a pattern emerging.*

### Conclusion to Chapter 19

- By now you should feel confident to tackle more or less any kind of assignment that involves putting together a presentation.
- You should have recognized that all presentations boil down to knowing your stuff, doing what the assignment instructs and communicating clearly.
- Unfortunately, different people will have different ideas about what an assignment requires and how best to achieve it. Your ultimate challenge, therefore, will be to apply your presentational skills when working as part of a group. We'll consider that challenge in the next chapter.

# 20

# How do I work as part of a group?


## Chapter summary

**Group work is an opportunity to develop your teamworking skills. The key stages are familiarization, planning, doing the work and delivering the final product. You need to know the members of your group, understand what you have to do, organise the division of tasks and a timetable, keep a check on progress and allow time to bring it all together in the final product. Use the experience of group work to reflect on what you are good at.**


## Why will I be asked to do group work?

Traditionally, group work has been most common in fieldwork, where there is a task to be performed and it needs more than one person to do it. For example, if you are carrying out a surveying exercise, it's usually physically impossible to do it on your own. Also, there are issues of safety. It is safer to have groups of students working together in the field than have students out on their own. Even if it's possible for one person to do the task on their own and there are no safety issues, the limited time available on a field course often means that collecting a reasonable amount of data to address the problem in hand is too big a job for one person. Similarly, it might be useful for you, individually, to learn how to conduct a few questionnaires; but, after you've done a couple of dozen, diminishing returns set in. There's not much to be learnt from questionnaires 50 to 100, although you may need 100 replies to get sufficient data for statistical analysis. So, having five students doing 20 questionnaires each achieves the educational goals, gets the data and doesn't overload any one student. Group work works!

In recent years, group work has increasingly been introduced to non-fieldwork classes. One reason for this is that it is considered good training (development of a transferable skill) for you to learn to work in a group. In a job after you leave university you are likely to find yourself part of a team, so learning to work with other students gives you useful experience. It will teach you about the issues to be faced in working in a group, and may, if you reflect on the task, tell you something about yourself. How well did you cope with working with others? Did you lead or follow? A second reason is that tutors are always looking for more diverse ways of identifying strengths of particular students, and project work is becoming an increasingly popular assessment method. In project work many of the issues that we identified for fieldwork are relevant (time constraints, need for a sufficient amount of data, etc.) so group work is appropriate.

## What stages does a group go through in working together?

When we talked about getting started in writing an essay, we said there were several stages to go through. It's the same with group work. There are several stages to working in a group successfully.

### 1. Familiarization

There are two aspects to familiarization: getting to know each other and getting to know the task. First, you may or may not know all the members of your group, so getting to know everybody and what they can bring to the group is an important first step. Have an informal meeting over coffee. If you don't all know each other, it's a good idea for each of you to say a few words about yourself (you may have done this sort of thing in first-year tutorials).

Once you know each other, you can move on to the second element of familiarization: getting to know the task. What have you got to do? What will you have to deliver at the end? How long do you have for all of this? How will you be assessed? Make sure you have clear answers to all of these questions. Write them down and circulate the information to the entire group.

### 2. Planning

Once you know what you have to do, you can move on to working out how you will do it. Divide the task into its component parts. These will vary, depending on the task. For example, if the task is to produce a poster, you may need to gather background information, decide what is to go on the poster, write the text, produce the illustrations and put it all together.

**Make sure that everybody in the group has a say in the brainstorming session where you put your ideas together.**

Other tasks may involve data collection and analysis, as well. Draw up a flow diagram of the task that you are all happy with. Give everybody a copy of the flow diagram.

Decide who will do what. In almost all cases, you will find you don't need the whole group to be involved in doing everything. Then produce a timetable.

Remember that you can have some tasks going on simultaneously. The important points about the timetable are: (i) to identify crucial points when there will be a hold-up if certain things are not done; and (ii) to allow more time than you think for bringing it all together at the end. Allow some time at the end for perfecting the final product (see Chapter 15).

## 3. Doing the work

Now that all members of your group know what's to be done and how you're going to do it, it's time to leave the coffee shop and head off in your separate ways to the library, the field, the laboratory, etc. and get on with it. But before you go, decide on a programme of meetings or e-mail contacts.

**Don't assume that everything will go smoothly until the next step: ensure that it does.**

So, if you are doing a poster and you are all off to collect data and you have agreed to complete this within a week, don't leave it until the week is up to get together again. At the very least agree a mid-point check-in by e-mail to confirm everything is on target. Fix a time for a meeting, if you need one. You can always cancel it if you don't, but organizing one at short notice is more difficult. If the project is a longer one, arrange weekly meetings as a matter of routine to report on progress and iron out any problems. Take notes at these meetings and circulate them. They are useful to make sure everybody knows what's going on. If disaster occurs and your group work falls apart because somebody failed to do their bit, the notes will be useful to your tutor in deciding what to do about it. It may sound a bit tough, but the chances are the group work will count towards your degree assessment. You don't want to be penalized because one member of your group didn't deliver.

## 4. Delivering the final product

Your group work will have to deliver some final product to be assessed. It could be a talk, a poster, or any of the other things we've discussed in this book. You might want to review the relevant chapters for your particular assignment. There's one big difference this time, which is that you are not now producing these things on your own. So how do you deal with this?

If you have to produce an essay or write a report, don't sit down together and write it as a team. It won't work. Even two people writing something together have to have a lot of practice for it to be successful. It would be better to divide the report into sections and each take responsibility for drafting one section. If you agree to write sections individually, you will need to allow time for editing your drafts together. We've written this book together in several stages. We started by writing our designated chapters separately, then we added bits to what the other had written in each chapter, then we went through the whole thing to impose a consistent style, and then we both read it all to make sure we thought it was OK. You need to do the same.

If you're giving a talk, make sure you have a common 'script' to which you each work. A piece of group work should still appear as if it is written or spoken by one person.

Eric says: *Why is a good group presentation like a failed coal mine? Because it is a seamless whole.*

## What if I end up in a group with Billy Thick and Dumb Doris?

The way of deciding who goes in which group varies a lot, even with the same tutor. Sometimes you will be allowed to select your own group, sometimes the tutor will choose who is in which group, and sometimes it will be done by drawing names out of a hat. Whichever way is used, you could end up in a group of Billy Thicks and on your way to a low mark, whereas you would like to be in a group of Susan Smarts and on your way to a high mark. How you deal with this, and how it affects the mark you get for your group work, will depend partly on how you handle the situation to manage the group, and partly on how your tutor assesses the work.

## Managing the group

One reason why doing group work is a useful transferable skill is that it's not just about getting the work done, but also about managing time and people, and about being organized. Writing an essay on your own, you can be pretty disorganized and still get away with it (though we don't recommend that approach). With a group, this is well nigh impossible – unless you are all disorganized in exactly the same way.

There are three elements to managing a group: managing time, managing the work and managing people.

### Managing time

Because you are all making different contributions to the group activity (or at least there are very likely several things being done at the same time), it is important to divide up the time into blocks in which certain things will be completed. At the planning stage you will only be able to make an estimate of how long each task will take. Allow for the fact that some things may take longer than expected. As we've already said, you should make a timetable. Include in the timetable a schedule of meetings. These should occur at times when a set of tasks is due to be completed. Make sure you build in some spare time. It's very important that the last bits of the work don't end up rushed because early parts overran their allotted time. Almost certainly, it's what's done in the last stages that will figure prominently in whatever way you are assessed. You can't say to your tutor, 'We know the presentation was rubbish, but you should have seen our soil pits.'

## Managing the work

Effective division of the work will make the most of the time available and guarantee the group achieves the most in that time. So think about how much effort needs to go into each activity. At the planning stage, when you divide up the task into its component parts, decide how much effort needs to go into each bit. Write down the number of person-days to be allocated. You can then decide if this can all be done by one person working for $x$ days, or by three people working for $x/3$ days.

An important part of managing the work is making sure everybody knows what is to be done. So always take notes at your meetings and circulate them afterwards. The notes should include who has agreed to do what and by when. Each meeting should start with a review of the actions agreed at the last meeting. Ideally, everything will have been completed and this review will consist of a series of successful reports. But it may not. There may have been unforeseen problems, or one of your group may be just lazy. If Eric hasn't completed his task because he couldn't be bothered to get up that day, record that in the notes of the meeting. At best it will encourage him to improve; at worst it will be useful when you want to justify why he should receive a lower mark than the rest of you.

## Managing people

The most obvious person in the group is often the group leader – the one who says most in the meetings and who takes the lead in making decisions about the work. It's easy to think this person is doing more work than the rest of you. But don't jump to conclusions. The member of the group who doesn't say very much in meetings, who accepts whatever task is assigned by the group, but who always turn up at the next meeting with everything completed and with copies for everyone, plays just as important a role. There are many aspects of group work and they all have to be done successfully for the group to achieve its goals. A successful group will be one that recognizes all the things that need to be done and exploits the strengths of each of its members. An unsuccessful one will be one where everybody wants to lead the group, but nobody wants to do the tedious tasks. In an ideal world, the members of the group would know each other well enough to know what each was good at, and divide up the tasks accordingly. But you are not likely to be in that situation. You can, however, help the group get close if you've thought about your own strengths and weaknesses beforehand.

# Know yourself before you start

What things are you good at? In general, we enjoy the things we're good at, and put off the things that we don't do well. So think about how you have approached tasks when you've been working on your own. Do you spend ages in the library reading far more than you need, and then find you've left too little time for writing it all up? Do you rush into writing the report and then find you've not got a vital piece of information and have to go back and get it? Are your essays criticized for having a lot of detail, but lacking a critical overview? Or are they criticized for having a good structure but lacking detail? Are your reference lists often incomplete? In this book, we have tried to equip you to cover all of the aspects of your coursework with equal effectiveness. But you will know that some bits you find easier and enjoy more than others. In group work you have the opportunity to exploit your strengths and get somebody else to cover the bits you are less good at. But you can only do this if you know these things. It's unlikely that the rest of the group will know. So if you are asked to work in a group, take some time before the first meeting of the group to reflect on these issues, and go along to that meeting with some idea of the aspects of the work that you feel you could do best. Then, when the group is looking for volunteers to take on a particular task, you can volunteer for it confident that you will do a good job.

# How will group work be assessed?

Because tutors recognize the value of group work, but want to make sure that the assessment is fair to individuals, a lot of effort is put into giving each student a mark that reflects his or her real contribution to the group. You will almost certainly find quite a wide range of schemes in use, even within one department.

**Nobody has cracked the problem of finding a way to ensure that the individual marks awarded to each student within a group are completely fair.**

One way to find out how much any one person has contributed to the work of the group is to ask the group. This is known as peer assessment. An example of this approach to determining the individual mark that a student gets for group work is shown in Box 20.1.

The scores are chosen to prevent students all giving each other a bonus – a member who made a major contribution throughout would receive the group grade, but one who contributed little in all categories would receive the group grade minus 16 marks. It may be desirable to negotiate the criteria, or even the penalties.

## Box 20.1 Peer assessment of contribution to group projects

*Each student receives the same group mark subject to the following adjustments.*
*Each student is required to mark every other group member anonymously.*

Module No ............................. Project.........................................................

Student (name) ......................... has contributed to the group work as follows:

| | Major | Average | Little |
|---|---|---|---|
| Leadership and management | 0 | –1 | –2 |
| Ideas and suggestions | 0 | –1 | –2 |
| Data gathering | 0 | –1 | –2 |
| Data analysis | 0 | –1 | –2 |
| Report writing/poster preparation | 0 | –2 | –4 |
| Verbal presentation | 0 | –2 | –4 |
| TOTAL DEDUCTION | | | |

A second approach is to ask groups to keep a logbook of their activity so that the tutor can see who has done what, and adjust marks accordingly. Often a tutor will ask to see the minutes of your meetings to check that any peer assessment is fair.

Check with your tutor at the start of a piece of group work how it will be assessed, how the marks will be distributed among the members of the group, and what you will need to provide for the assessment to be completed. Make sure your group can produce everything that is required.

# Learn about yourself after you've finished

We've said that group work is popular with tutors because it provides you with a transferable skill that you are likely to need later in life. Once you have worked in a group, it is useful to reflect on what you learnt about yourself. Did you say too much at the group meetings? Did you find you left meetings dissatisfied that the work was being planned in a particular way, but that you had been unable to get across why you thought this was wrong? Did you enjoy the experience of working with others? Did you learn that others were better/worse at some things than you are? Did you always do what was required of you to the best of your ability? Did you let down your colleagues? Did you find other members of the group annoying? Try answering each of these questions as honestly as you can. Remember the answers for the next time you work in a group.

# Further information

A useful web site to look at for some further information on group work is http://bokcenter.fas.harvard.edu/docs/wigintro.html

## Conclusion to Chapter 20

- Group work will help you learn how to operate effectively as part of a team.
- There are stages in effective group work – just as for essays.
- Keep everybody informed of what is going on.
- Keep a record of the group's activities.
- Divide up tasks among the group: don't all try to do everything.
- Afterwards, think about how you performed as part of the group.

# 21

# Help! It's all going wrong, what can I do?

## Chapter summary

A lot can go wrong when you are writing an essay, doing other coursework, or sitting an exam. There are a few problems that are very difficult to solve, and must be avoided at all costs. However, most problems can be solved as long as you recognize them. Careful preparation will prevent most serious problems, and here we offer some advice to help you deal with those that you haven't managed to avoid.

## What can go wrong?

Everything can go wrong.

- Things can interrupt you and make you late.
- Accidents can destroy the work you've done.
- Mistakes can make you do the wrong work.
- Carelessness, laziness and poor planning can make you do a bad job.
- Things can break or disappear when you need them.
- Jobs can take longer than you expected.
- When you read through your work you can discover it's rubbish.
- The work can just be too difficult for you.
- Bad things can happen to good people.
- Problems that we've never even dreamed of can arise.

There are more things that can go wrong than there are things that you knew were even involved in completing an assignment. Look at the list of excuses that students have given for submitting their essays late (Box 21.1) and you'll get an idea of the range of things we are talking about.

**Box 21.1 Excuses that students have given us for submitting their essays late**

- I thought the deadline was next week.
- I thought we could choose *any* question from the list.
- All the books were out of the library.
- I had three other essays due the same day.
- I broke my leg in 19 places.
- The cat died.
- The car broke down.
- The printer ran out of ink.
- The computer destroyed my essay file.
- The dog ate my printout.
- My parents made me go on a skiing holiday.
- I had to take my friend to hospital.
- I had 'food poisoning' – that last pint was a bit off.
- I just couldn't think what to write.
- My introduction didn't match my conclusion when I'd finished.
- I couldn't find an example of 'x'.

## Will the tutor make allowances for me when things go wrong?

If you look at the list of excuses in Box 21.1, you will see that a tutor could dismiss them all by using a small stock of standard responses. Try matching these responses to the excuses we listed and see how many of the excuses survive.

- You should have checked.
- You should have had a back-up.
- You should have been more careful.
- You should have worked harder.
- You should have left more time in case of problems like that.
- You should have asked for help earlier.

None of the excuses in Box 21.1 can get through these defences. Let this be a lesson to you.

**Tutors will expect you to have taken precautions to avoid things going wrong and to have made preparations to deal with the situation if things do go wrong.**

If you are not prepared when you should have been, it is only fair that you are penalized relative to students who *were* prepared. Being prepared is part of the essay-writing game. Tutors will make allowances only in very special circumstances when you cannot be expected to have been prepared for the problems that arose. For example, you may be granted an extension to the deadline or you may be given a few extra marks if your performance can be shown to have been affected by something like the death of a relative, a major domestic upheaval such as a divorce, or a serious illness. In such cases, however, the examiners will take careful note of the timing of the event and will ask themselves whether they can apply one of their standard responses. It's hard to avoid 'You should have left more time in case of problems like that', but in genuine cases of serious bad luck your problems will be taken into account as long as they are backed up with documentary evidence like a doctor's note. Later in this chapter we'll look at what you should do if you face a serious problem like this.

## How can I reduce the likelihood of things going wrong?

Most problems are avoidable, and good working practices minimize the risk of things going wrong.

**Good students always seem to have less bad luck than bad students do.**

In every group there are one or two students who you just know are going to have things go wrong for them. They think it's because they are unlucky, and they will tell you that they've always been unlucky. In fact, the reason things always go wrong for these students more than for the others is that they create the opportunity for bad luck to do more serious damage to their essays than the same luck would do to the essays of better prepared students. They leave themselves vulnerable to attack. They don't take precautions to defend against the little misfortunes that are likely to strike any of us.

There are a number of sensible precautions that you can take.

- Start early.
- Plan carefully.
- Work hard.
- Learn from feedback on previous assignments.
- Check everything repeatedly.
- Tell people about what you are doing and see how they react.
- Keep back-up copies of everything.
- Leave lots of spare time at the end.

If you follow these rules, things are less likely to go wrong than if you don't; if things do go wrong you are more likely to spot them early before any real harm is done; and if something serious goes wrong you will be in a stronger position to find and implement a solution. If the worst comes to the very worst, you will at least have the sympathy, support and understanding of your tutor. However, if your 'misfortune' stems from bad planning, carelessness, laziness, not listening to advice, not using common sense and not giving yourself enough time to write the essay in between your social activities, your tutor will be less sympathetic.

*Eric's friend Billy Thick says:* *My tutor never lets me get away with anything. I'm just unlucky.*

## What should I do when things do go wrong?

*Eric's friend Clever Catherine says:* *Fix 'em.*

When something goes wrong, don't be surprised. It is only to be expected. It's part of the game. When you do an assignment you are being tested on: your knowledge and understanding of the subject; your ability to communicate effectively; and your ability to deal with problems when they arise. Sorting things out when something goes wrong is just part of the job. Don't worry. Don't panic. Oh, and by the way, just in case any of you might have thought about it, don't try cheating. If you get caught it will destroy your whole degree, will go into all your job references, and will mark you out for the rest of your life. Cheating is in fact the worst thing that can go wrong and one of the trickiest problems to solve. More on that later, but for now, what *do* you do?

To start with, make sure you correctly identify exactly what has gone wrong. There's no point setting out to solve problem 29B (can't think of good conclusion to essay) if you are actually facing problem 2A (don't know answer to question). What *is* your problem?

Once you recognize your problem, you need to decide whether you can fix it yourself or need to go and see somebody. If you do need to go and see somebody (your tutor, a counsellor, a doctor) then you should do so sooner rather than later. Look at the section on 'Asking for help' below. If you can fix it yourself, then get on with it and that will be you sorted! Most problems can be fixed by a bit of careful thought, reading or re-writing. If your specific problem is in the list we go through later in this chapter, we can give you some specific advice. If not, we can at least give you some general pointers that you can apply to your particular situation.

## Are there problems that can't be solved and must be avoided?

On the whole, the only problems that can't be solved are the ones that escape being noticed till it's too late. If you spot a problem in time, you can fix it. However, some problems do take a little time to fix, and if you notice them 30 seconds before the end of the exam, it's a bit late. Some problems call for solutions in the 'go back and start again' category. Unless you have time to do that, these must be avoided. Take a look at the nightmare scenarios in Box 21.2.

### Box 21.2 Nightmare scenarios: problems that are hard to fix and are best avoided

*Scenario 1: Misunderstanding or misreading the exam instructions.*

I was just coming to the end of the exam, with about 2 minutes to go. I had finished all three of my essays and I was just waiting for the finishing bell. To pass the time I was inking in the spaces in every letter 'o' on the question paper. I was happy until I got to the o in 'four' where it said 'You must answer four questions'. Just as I realized what it meant, the bell went.

*Tutor's sympathetic response:*

You should have checked. You should have been more careful. You should have left more time in case of problems like that. You're a prat.

*Scenario 2: Misinterpreting or misreading the question.*

Walking into the office to hand in the essay I was talking to Jill about how I'd used Snowdonia as an example of a glaciated area in my essay. Jill said she didn't think there were glaciers in Snowdonia, but I said that there were in the past. Jill got out her essay and we looked at the title: 'Discuss the impact of glaciers on fluvial systems in modern glaciated landscapes'. Now I'm worried that I should have used examples from landscapes that actually have glaciers in them today, rather than landscapes that used to have glaciers in them. I think I misinterpreted the question.

*Tutor's sympathetic response:*

You misinterpreted the question. You should have checked. You should have been more careful. You should have talked to people about what you were doing. It's very unfortunate: 36%, D.

The problems that must be avoided are generally quite simple to avoid, and even if you can't avoid them they are easy enough to fix if you just have enough time to go back and do the work again. The problems in Box 21.2 would not be serious if they were identified a week before the deadline. That's why your tutor will not have much sympathy if you get caught out. However, we are much more sympathetic than most tutors, so we will give you some advice about what to do if things really start to go wrong.

# Solutions to specific problems: what should I do if . . .?

## I can't get started

Take little steps and begin with something small and straightforward. Copy out the question. Re-order the words until they look like an answer, so 'What causes X?' becomes 'X is caused by . . .'. Look again at Chapter 6. Go away and have a break, come back fresh and try again. Buckle down and sit there till you've got something written; even if it's weak it will be a start and you can improve it later. Just write anything to break into the clean white page. Browse through books and journals for inspiration. Ask friends how they have started. If you can't get started it's probably because you don't know the answer.

## I don't know the answer

Look it up. Think about it. Work it out. Ask somebody. The clue to the answer is always in the question. Re-order the words in the title until they look like an answer (see '*I can't get started*'). The answer to many questions in geography is 'It depends', so you can try to get at the answer by working out what it depends on. For example:

Q: To what extent can A be explained by B?

A: The extent to which A can be explained by B depends on climate. In warm climates B explains A completely. In cool climates A is not explained by B at all.

Q: Examine the factors that control X.

A: The factors that control X depend on climate. In warm climates X is controlled largely by A and B. In cool climates X is controlled mainly by Y and Z.

What does your answer depend on? Would your answer be different for different climates, different points in history, different planets? If so, that gives you a way into the answer and also a way into developing an interesting plan. If you still can't think of an answer, you probably don't know enough about the topic.

## I don't know enough about the topic

If you are doing an exam, then Plan A should be to do a different question for which you do know something. If you are stuck with this question, then you will have to work around not knowing much, but you are unlikely to get a very good mark. If you don't have any information, you probably can't be very confident about your answer and you won't be able to provide much evidence to support it. You might try to dazzle the examiners with your bright ideas and hope they don't see the gap where the information should be, but it probably won't work – although you will get some credit for the bright ideas. Do what you can. Don't just give up – you will get credit for what you do manage to write. If you just give up and leave the exam, you will not get any benefit of the doubt. If you have a go, you may. If you are doing coursework, not knowing much is less of a problem unless the deadline is upon you. If the deadline is upon you, follow the advice we offered for exams, otherwise you can just go away and find the facts. Look them up. Not knowing enough need not be a permanent condition.

## I can't find any material

Look harder. Look in different places. Look for different material. Use different search or index terms. Look for material close to what you want and it may give you additional leads to the actual material you need. If you can't find 'Pingos in Finland' try looking for 'Pingos in Norway' instead. If that doesn't work, try Pingos in Scandinavia, in the Arctic, or just Pingos. If you can't find anything at first it might be because you are being too fussy. Try 'Periglacial landforms' or 'Arctic landscapes'. Information about pingos in Finland is out there somewhere. Phone the Finland Tourist Board. Have you looked in relevant bibliographic search engines on the web? Ask your librarian for help. Look in the reference lists at the backs of textbooks. Look in reference lists at the ends of papers. The material is out there.

## My plan is too complicated

Complex issues can generate complex essays that will have complex plans, but you need to communicate clearly. Make sure that you are not including superfluous material. Are you focusing on the key points and answering the question directly? If your plan is a mess, you probably need to group separate items together into bigger units. Put rings around sets of points that can be handled together. If your plan is web-like, you should turn it into a more linear or hierarchical structure to enable you to turn it into essay text. Identify the *most* important points and put them at the top. If you can't escape the complicated web structure, put that structure as a figure in the essay to help to make it clear to readers. Clear signposting in the essay will help to keep the reader on track.

## My plan is too simple

Are you sure it's too simple? Sometimes simplicity can be elegant. You don't want to make your plan unnecessarily complicated. As long as you make all the points that you need to, simplicity is a virtue. If you really need more complexity in your plan try adding an alternative viewpoint. Try seeing if any of the arrows on your flow chart should really be double-headed. Try reversing some of your relationships. Try drawing a circle around your whole plan, drawing a big arrow pointing at the circle from the left-hand side of the page, and thinking what should go at the other end of that arrow. What controls everything that you have so far included in the plan? This will give you an extra layer of material to talk about: an additional level in the hierarchy of your plan. Read more. Think more. Get more ideas.

## My essay is far too long

The most common cause of over-long essays is verbosity. *Go through your essay and take three words out of every sentence that runs to 15 words or more.* That last sentence had 19 words, so we should reduce it by three to 16 words: *Remove three words from every sentence in your essay that runs to 15 words or more.* That's 16 words, which is still more than 15, so we need to remove three more, which would make 13: *Remove three words from sentences more than 14 words long in your essay.* OK, that's not as pretty as the original, but it makes the same point quite clearly and it is about 30% shorter. Do that through the whole essay and see the difference! You can also save a lot of words by putting information into diagrams instead of text. Data can go into tables. Descriptions can go into maps and drawings. Relationships can be shown graphically rather than in text. If you are still too long, think about whether you repeat yourself unnecessarily. Do you have two examples where one would be adequate? Do you make the same point twice in one paragraph? Finally you can go through and remove all the meaningless phrases like 'Thus it can be seen that' and 'My own opinion is that', the meaning of which would be clear even if the words weren't there.

Eric says: *If your essay is too long, use fewer words.*

## My essay is far too short

Make it longer. If your essay is too short it is probably because your plan is too simple or because you don't know enough about the topic. Find out more about the topic. Develop your plan. Insert an additional case study. Think of an alternative point of view and make it into an interesting penultimate paragraph (Chapter 11). Make sure your essay has all the sections that a good essay should have (Chapter 7). Read more and get some extra ideas. Make your essay longer only by increasing the worthwhile content, not by adding useless words. In an exam, where you can't go out and get more facts, try adding a paragraph that begins with: 'An alternative approach to this topic . . .'.

Eric says: *Only use more words if it enables you to say more things.*

## My essay drifts off the point

This is a problem that should be avoided by careful planning. However, if you discover the problem when reading through the finished essay it is a bit late for planning. You should take remedial action. In a coursework essay, if you have time, go back and re-plan. If your plan drifts off the point, fix it. If your plan is good, fix the bits of text that drift away from it. Take out the sections of text that lead away from the point and replace them with relevant material. In an exam, you might want to cross out sections that flatly ignore the title and write 'irrelevant' in the margin. It will save the examiner a job and demonstrate at least that you recognize your mistake. Write new text that explains what the real point is and provides relevant material. If this won't fit in the original space, you can as a last resort put a note in the text saying 'See additional material at end of essay' and then put the extra material, clearly labelled, at the end. We don't recommend this approach except as a last resort. Better planning is the better solution. Tacking a bit on at the end may get you a few extra marks, but better planning would have got you a lot more. In a coursework essay you should make these repairs 'invisibly', so that the finished essay shows no sign of ever having been off track.

## I am running out of time

This is another one that could have been avoided with careful planning. You need to make more time or work faster. Give up some extracurricular activities. Reschedule your other commitments. Stop messing about and get on with your essay. If you have a really good excuse, see your tutor and you might get an extension. If you don't have a really good excuse, go and see your tutor anyway. You never know your luck. If you are seriously short of time (for example, in an exam with 3 minutes left) focus on the essentials. Remember the main things that the examiners are looking for: knowledge, understanding, an answer to the question and communication skills. Put down your remaining material in note form as bullet points. Jump to the conclusion and summarize your answer. Make it clear that you have run out of time, not out of ideas: you will lose marks, but you may lose fewer for bad timing than you will for ignorance.

## I've missed the deadline

Go and see your tutor. Don't just ignore the problem. There may be a get-out clause somewhere and your tutor may be able to help you find it. The longer you wait before seeking help, the less likely it is that you will find it. When you go to see your tutor, take along whatever work you have already done and a plan for catching up with the work you have not done. This will show that you are at least making an effort. Asking to have your deadline extended until tomorrow at 3 pm so that you can attend a funeral today and do the work tomorrow morning is better than asking for an unspecified extension to do the work 'soon'. Show that you have a plausible recovery schedule.

## My printer has run out of ink (and other technical setbacks)

This is only likely to be a serious problem if you have left printing to the very last minute before the deadline. That was a mistake. Always aim to print the essay out a day or two before the deadline, so if the ink runs out you have time to get more. That technique of problem avoidance applies to most technical problems. Can you borrow a friend's printer? Will your tutor accept an electronic submission at the deadline followed by an identical print-out very shortly afterwards? Will you have to copy out the whole thing by hand and submit a handwritten version? That will teach you not to leave things till the last minute!

Eric says: *I thought you guys were supposed to be sympathetic.*

## I have broken my leg in nine places

Good excuse. Nice one. Go and see your tutor – there's a good chance of getting an extension. On the other hand, do you actually type with that leg? Shouldn't you already have the bulk of the essay done by now? Even with a really serious problem like this, don't expect everything to fall into place for you without some work on your part. Don't forget to tell the department that it was because of the broken leg that your essay was late. Don't forget to get a copy of the medical certificate to the office. Make sure that your tutor has a record of your injury and remembers to bring it to the attention of the end-of-year exam board. Get well soon, but try to be more careful in future.

## I'm giving a talk and I've gone blank

Nightmare. You're stood up in front of the class, halfway through saying something, and you've lost it. Suddenly you feel hot. Seconds pass. Plan A: take a moment to look at your notes, find your place, and resume normal service. Don't worry about the slight delay: the audience will only mind if you look flustered. If you stay relaxed, so will they. Plan B: your notes are rubbish and they don't help. Say 'Let's move on to the next point', abandon the section that you have got lost in, and start the next section. Plan C: you don't even know what section you are in. Say 'Let's remind ourselves of where we are' and put up your first OHP or slide, which sets out the structure of your talk. If this doesn't remind you where you are, figure out how long you still have left to talk, figure out how far down the list you would need to be by now in order to finish on time, and just start again at that point on the list. Generally an audience only knows that something has gone wrong if you tell them (by panicking). If you stay calm each person in the audience will assume it's their fault that they got lost. They are unlikely to compare notes afterwards and discover it was your fault. (Of course, your tutor will almost certainly assume it was!) Don't be afraid to pause and think. You can extend your pause by taking a drink of water, writing on the board, adjusting the lights. As a last resort there's always a calm and impressive: 'I'm sorry but I seem to have lost my place. Let's take a break for a moment while I sort out my notes.'

## My brain has gone numb

If this seems to be a temporary condition, you should take a break. In an exam you can just sit back and think of something else for a few minutes. In coursework you can go out for a walk and a cup of tea with your neighbour. Try doing a different piece of work for a bit, and come back to this one when you're fresh. Talk about the assignment with other students. You can reduce the chances of suffering from brain numbness by being a well-balanced and healthy individual who gets plenty of sleep at night, eats well and exercises regularly. Fish is supposed to be good for the brain. So is oxygen. Take a brisk walk and eat a tin of mackerel.

## I don't know when to stop

Stop at the end. Stop when you have finished. Don't ramble on after you've done the job as we are doing with this sentence. Knowing when to stop is easy if you know what you need to say and can recognize what you have already said. The key to this is understanding the question and having a well-structured plan. If in doubt, stop sooner rather than later and certainly stop before you exceed the word limit for the assignment. In a talk, make sure you make your point and finish before the audience gets bored.

## I'm thick

No thicker, probably, than many students who have achieved perfectly respectable grades. Being thick just means you have to work harder. It's tough, but that's the way the world works. Some bright sparks just sail through their assignments without a care. Most of us have to slog away a little. If you think you're thick, let that motivate you to work harder so that you do just as well as the bright buggers. Your tutors will help you. If you are willing to work hard, they will be willing to work hard with you to help you to do as well as you possibly can. Read this book (you are doing, well done), follow our advice, try hard and you will do OK. Sometimes people who are a bit thick are tempted to cheat. If you are tempted to cheat, read the next section.

## I am tempted to cheat

Don't do it. No single piece of coursework and no single exam essay is likely to change your life, unless you cheat, and get caught, in which case it will haunt you forever. If you cheat you will score zero for the assignment, you will almost certainly forfeit the module, you will probably have to re-sit the whole year, you may be expelled from the course, your record will be marked for life, referees who write in support of your job applications will have to take your cheating into account, and almost everybody you meet will think you are scum. If you are tempted to cheat it is probably either because you are lazy or because you have some problem with the work. If you are lazy, look at the 'bare minimum' section in Emergency rescue on the next page. If you have some other problem so serious that it tempts you to cheat, go and seek help. More or less anything is better than cheating: even having to go and talk to your tutor about being tempted to cheat!

## I have cheated

Uh oh. Come out of the building with your hands above your head. Seriously, this is just about the worst thing that could possibly have happened. You need to fix it, quick. If you have not yet submitted the work, erase everything that was derived from cheating and rewrite the essay using completely new material. Better still, if possible, write a whole new essay. Then go and see your tutor and explain what you've done. Better to come clean than to try to hide it and have it come out later as a scandal. If you have already submitted the work, or if the cheating can't be undone, go immediately and see your tutor. Explain what you've done, explain that you realize and regret your mistake, ask to have the submission disregarded and ask to be allowed to resubmit a new essay. This will certainly entail a severe marks deduction, and you may even score zero, but that is better than being caught cheating.

### I have been caught cheating

There comes a point where it is too late for us to help you. This is that point.

## Asking for help

Don't be afraid to ask for help. You can ask your tutors, you can ask other staff in the administrative and student support sections of your institution, you can ask your doctor, your lawyer, your family and friends. Help is all over the place. Your tutors are there to help you learn, and answering your questions is one of their jobs.

**When you ask for help you will probably get a much more sympathetic and constructive response if you show that you have already made some effort to help yourself.**

If you go to the tutor 5 minutes after the essay has been set and say, 'I'm stuck, how should I begin?', the tutor will surely be tempted to say, 'Don't be so lazy, think about it, try, use your experience and intelligence, work on it for a while, get your brain in gear, pull your socks up.' On the other hand, if you go back after a couple of days with two or three alternative plans for how you might tackle the essay, with some references to relevant papers, and with a thoughtful question about how best to move forward with the essay, your tutor is more likely to say, 'So far so good, try this, change that, add these . . .'.

When you are struck by a real disaster such as a medical, personal or domestic crisis, seek help right away. If you seek help from someone outside your department, such as a doctor or counsellor, you should let the department know. The examiners will not be able to make allowances for problems that they don't know about. The same applies to any long-term conditions or disabilities such as dyslexia. If they don't know about it they can't take it into account.

## Emergency rescue: a disaster kit – the last resort

If you need this section you are in real trouble. This is like the inflatable life-raft in a helicopter: you don't inflate it unless the sea water is lapping around your ears. This really is the last resort. If you are going to score zero, we can improve on that for you, but if you are already scoring a decent pass you don't need to read the next few paragraphs. If you are a tutor checking what advice we are giving to your students, you should probably cover your eyes. If you are really sinking, read on.

## What's the bare minimum I can do?

Not everybody wants to do well. Some people just want to get by. Some people want to do the bare minimum that will allow them to scrape through the course. This is a very dangerous game to play, because if you slip below your target you will fail the course. Aiming high is much safer. However, if you want an easy ride and don't care too much about the grades, there is a bare minimum that will get you through. If you are exceptionally bright you can get away with doing less work than if you are not, but either way you need to achieve the same goal. That goal is explained in the qualitative assessment guides that we've referred to earlier in the book. Previously we were focusing on the requirements for a good grade, but if you want to risk it you could just look at the requirements for a bare pass, and aim for that. Box 21.3 identifies key points from the bottom end of the assessment guide that we gave in Box 4.3. You could think of these as 'minimum requirements'.

### Box 21.3 Key points from the bottom end of the assessment guide that we gave in Box 4.3 (you could think of these as 'minimum requirements')

Low 3rd-class honours (40–44%)

- Must provide a relevant answer.
- Must show some knowledge and understanding of the subject.
- Must recognize some of the main issues raised by the assignment.
- May include serious errors, omissions or irrelevant material.
- May lack any evidence of independent thought.
- May lack evidence for arguments.
- May show no indication of independent reading.
- May not be presented to a high standard in the style appropriate to the assignment.

Pass without honours (35–39%)

- Must show some understanding of the relevant issues.
- Must show a basic appreciation of the style appropriate to the assignment.

Fail (< 35%)

- May lack knowledge and understanding of the subject.
- May be seriously flawed with errors, omissions and irrelevant material.
- May not be presented in a style appropriate to the assignment.

## What are the good excuses that will really get me out of trouble?

On the whole, you can't avoid doing your assignments, but if you are in real trouble you might either get an extension on the deadline or get some allowance for poor work or late submission when the examiners finalize your marks. At the end of the year the examiners will look at the final scores achieved by each student and consider adjusting them in light of any special circumstances.

If any of these special circumstances apply to you, write to the department and ask for special consideration:

- death of close relative;
- serious injury, illness or personal trauma;
- jury service, imprisonment or conscription for National Service.

However, you will only get special consideration if these circumstances affected you at just such a time that you could not have worked around them, and only if you have some form of documentary evidence to support your claim. For example, jury service is not an excuse if you were given 2 months notice of it: you could have arranged your work schedule around it. Death of a close relative is only going to be taken into consideration if it occurred before the deadline. Your own sickness or injury will only be taken into account if the examiners have an official medical certificate (doctor's note) in their file. Of course, none of these circumstances will be taken into account if you don't tell the department about them. Keep your tutor informed.

## Can I make up a fake excuse?

If your excuse turns out to be fake, your marks will turn to dust. If you want to try a fake excuse you'd better make it convincing. This would come under the general heading of cheating, and you know what happens if you get caught doing that (see sections earlier in this chapter). It will be a life-changing experience.

## Conclusion to Chapter 21

- By being careful and well prepared, you can minimize the chances of things going wrong and you can put yourself in a good position to deal with problems if they do arise.
- Problems can be solved. Some are solved easily, and we have discussed some of these in this chapter. Others are harder to solve, and no doubt some of you will have problems that we have not anticipated here. Seek help. Talk to your tutor. Don't keep it to yourself. There are people who will help you to sort things out.
- If in doubt, consult your tutor.
- Good luck.

# 22 A final word

Dear Reader,
In writing this book we have often felt as if we are stating the obvious. If you know how to do exams, essays and other coursework, then a lot of what we have written *should* be obvious. We have written the book for people who don't know how, or who are not very good at it. We have given advice that we hope will help those people to get better and to reach a decent standard.

To some extent we've offered a 'recipe' for cooking essays. We have identified the key ingredients and suggested how they should be put together. However, following a recipe is only the first step in learning how to cook. If you stick religiously to our recipe you will produce a basically competent product. It is up to you to breathe life into your essay and make it stand out from the mass-produced crowd. Good cooks modify recipes to reflect their own personality. When you become a really great cook you no longer need a recipe.

Between us we mark about 1000 essays each academic year. Over the years that we've been teaching that adds up to something like 40 000 essays between us. What we have written in this book is based on the experiences of seeing what those 40 000 essays have done well or done badly. It's sad to say, but most essays seem to do something badly. A lot of essays do a lot of bad things. If you have read this book, and if you are about to hand in an essay, we hope that we have helped you to do a little better than you would otherwise have done.

Good luck.

Peter and Tony

# Index